Advances in Intelligent Systems and Computing

Volume 1083

The series "Advances in Intelligent Systems and Computing" contains publications on theory, applications, and design methods of Intelligent Systems and Intelligent Computing. Virtually all disciplines such as engineering, natural sciences, computer and information science, ICT, economics, business, e-commerce, environment, healthcare, life science are covered. The list of topics spans all the areas of modern intelligent systems and computing such as: computational intelligence, soft computing including neural networks, fuzzy systems, evolutionary computing and the fusion of these paradigms, social intelligence, ambient intelligence, computational neuroscience, artificial life, virtual worlds and society, cognitive science and systems, Perception and Vision, DNA and immune based systems, self-organizing and adaptive systems, e-Learning and teaching, human-centered and human-centric computing, recommender systems, intelligent control, robotics and mechatronics including human-machine teaming, knowledge-based paradigms, learning paradigms, machine ethics, intelligent data analysis, knowledge management, intelligent agents, intelligent decision making and support, intelligent network security, trust management, interactive entertainment, Web intelligence and multimedia.

The publications within "Advances in Intelligent Systems and Computing" are primarily proceedings of important conferences, symposia and congresses. They cover significant recent developments in the field, both of a foundational and applicable character. An important characteristic feature of the series is the short publication time and world-wide distribution. This permits a rapid and broad dissemination of research results.

** Indexing: The books of this series are submitted to ISI Proceedings, EI-Compendex, DBLP, SCOPUS, Google Scholar and Springerlink **

More information about this series at http://www.springer.com/series/11156

Elżbieta Macioszek · Grzegorz Sierpiński
Editors

Modern Traffic Engineering in the System Approach to the Development of Traffic Networks

16th Scientific and Technical Conference
"Transport Systems. Theory and Practice 2019"
Selected Papers

 Springer

Editors
Elżbieta Macioszek
Faculty of Transport
Silesian University of Technology
Katowice, Poland

Grzegorz Sierpiński
Faculty of Transport
Silesian University of Technology
Katowice, Poland

ISSN 2194-5357   ISSN 2194-5365 (electronic)
Advances in Intelligent Systems and Computing
ISBN 978-3-030-34068-1   ISBN 978-3-030-34069-8 (eBook)
https://doi.org/10.1007/978-3-030-34069-8

This Springer imprint is published by the registered company Springer Nature Switzerland AG
The registered company address is: Gewerbestrasse 11, 6330 Cham, Switzerland

Preface

One of the most important factors conditioning the economic development of each country is transport. The need for improving and developing an inherently cohesive and efficient transport system, integrated with the global system by increasing transport accessibility while improving the safety of traffic participants and the effectiveness of the transport sector at the same time, is one of the primary goals pursued by modern traffic engineering.

This publication, entitled *Modern Traffic Engineering in the System Approach to the Development of Traffic Networks*, provides an excellent opportunity to become familiar with the latest trends and achievements in the field of contemporary transport network as well as traffic engineering challenges and solutions. The book has been divided into four parts. They are:

- Part 1. Traffic Engineering as the Source of Decision in Transport Planning,
- Part 2. Modelling as Support in the Management of Traffic in Transport Networks,
- Part 3. Safety Issues in Transport—Human Factor, Applicable Procedures, Modern Technology,
- Part 4. Structure and Traffic Organization in Transport Systems.

The publication contains selected papers submitted to and presented at the 16th "Transport Systems. Theory and Practice" Scientific and Technical Conference, organized by the Department of Transport Systems and Traffic Engineering at the Faculty of Transport of the Silesian University of Technology, Katowice, Poland. The subjects addressed by the authors include the current problems related to the system approach to the development of transport networks. The publication provides numerous practical examples showcasing innovative solutions which set the trends for the development of transport networks and traffic engineering, while at the same time they constitute sources for decision-making in transport planning, structure and traffic organization in transport systems. These solutions have a significant impact on improving the efficiency of functioning of transport networks and system, but first and foremost they give priority to human well-being and health, traffic safety, sustainable development of transport systems as well as protection of natural environment. For this reason, some of the articles included in the publication also address traffic safety issues in transport, i.e. the human factor, applicable procedures and modern technology. Moreover, modelling tool issues as support in the management of traffic in transport were also included in presented volume.

We would like to make the most of this opportunity and express our gratitude to the authors for the papers they have submitted and their substantial contribution to a discourse on the modern traffic engineering from the system approach perspective, for discussing numerous challenges facing transport networks, systems as well as traffic engineering in the contemporary world, as well as for rendering the results of their research and scientific work available. We would also like to thank the reviewers for their insightful remarks and suggestions which have ensured the high quality of the publication.

Readers interested in the latest achievements and directions of traffic engineering development will find in the volume a comprehensive material presenting the results of scientific research, diverse insights and comments as well as new approaches and problem solutions. With the foregoing in mind, we wish all readers a successful reading.

September 2019   Elżbieta Macioszek
                                                                     Grzegorz Sierpiński

Organization

16th Scientific and Technical Conference "Transport Systems. Theory and Practice" (TSTP2019) is organized by the Department of Transport Systems and Traffic Engineering, Faculty of Transport, Silesian University of Technology, Poland.

Organizing Committee

Organizing Chair

Grzegorz Sierpiński   Silesian University of Technology, Poland

Members

Marcin Staniek	Silesian University of Technology, Poland
Renata Żochowska	Silesian University of Technology, Poland
Ireneusz Celiński	Silesian University of Technology, Poland
Grzegorz Karoń	Silesian University of Technology, Poland
Marcin J. Kłos	Silesian University of Technology, Poland
Krzysztof Krawiec	Silesian University of Technology, Poland
Aleksander Sobota	Silesian University of Technology, Poland
Barbara Borówka	Silesian University of Technology, Poland
Piotr Soczówka	Silesian University of Technology, Poland

The Conference Took Place Under the Honorary Patronage

Ministry of Infrastructure
Marshal of the Silesian Voivodeship

Scientific Committee

Stanisław Krawiec (Chairman)	Silesian University of Technology, Poland
Rahmi Akçelik	SIDRA SOLUTIONS, Australia
Tomasz Ambroziak	Warsaw University of Technology, Poland
Henryk Bałuch	The Railway Institute, Poland
Werner Brilon	Ruhr-University Bochum, Germany
Margarida Coelho	University of Aveiro, Portugal
Boris Davydov	Far Eastern State Transport University, Khabarovsk, Russia
Mehmet Dikmen	Baskent University, Turkey
Domokos Esztergár-Kiss	Budapest University of Technology and Economics, Hungary
Zoltán Fazekas	Institute for Computer Science and Control, Hungary
József Gál	University of Szeged, Hungary
Anna Granà	University of Palermo, Italy
Andrzej S. Grzelakowski	Gdynia Maritime University, Poland
Mehmet Serdar Güzel	Ankara University, Turkey
Józef Hansel	AGH University of Science and Technology, Cracow, Poland
Libor Ižvolt	University of Žilina, Slovakia
Marianna Jacyna	Warsaw University of Technology, Poland
Ilona Jacyna-Gołda	Warsaw University of Technology, Poland
Nan Kang	Tokyo University of Science, Japan
Jan Kempa	University of Technology and Life Sciences in Bydgoszcz, Poland
Michael Koniordos	Piraeus University of Applied Sciences, Greece
Bogusław Łazarz	Silesian University of Technology, Poland
Michal Maciejewski	Technical University Berlin, Germany
Elżbieta Macioszek	Silesian University of Technology, Poland
Ján Mandula	Technical University of Košice, Slovakia
Sylwester Markusik	Silesian University of Technology, Poland
Antonio Masegosa	IKERBASQUE Research Fellow at University of Deusto Bilbao, Spain
Agnieszka Merkisz-Guranowska	Poznań University of Technology, Poland
Anna Mężyk	Kazimierz Pulaski University of Technology and Humanities in Radom, Poland
Maria Michałowska	University of Economics in Katowice, Poland
Leszek Mindur	International School of Logistics and Transport in Wrocław, Poland
Maciej Mindur	Lublin University of Technology, Poland
Goran Mladenović	University of Belgrade, Serbia

Kai Nagel	Technical University Berlin, Germany
Piotr Niedzielski	University of Szczecin, Poland
Piotr Olszewski	Warsaw University of Technology, Poland
Enrique Onieva	Deusto Institute of Technology, University of Deusto Bilbao, Spain
Asier Perallos	Deusto Institute of Technology, University of Deusto Bilbao, Spain
Hrvoje Pilko	University of Zagreb, Croatia
Antonio Pratelli	University of Pisa, Italy
Dariusz Pyza	Warsaw University of Technology, Poland
Cesar Queiroz	World Bank Consultant (Former World Bank Highways Adviser), Washington, DC, USA
Andrzej Rudnicki	Cracow University of Technology, Poland
František Schlosser	University of Žilina, Slovakia
Grzegorz Sierpiński	Silesian University of Technology, Poland
Jacek Skorupski	Warsaw University of Technology, Poland
Aleksander Sładkowski	Silesian University of Technology, Poland
Wiesław Starowicz	Cracow University of Technology, Poland
Andrzej Szarata	Cracow University of Technology, Poland
Tomasz Szczuraszek	University of Technology and Life Sciences in Bydgoszcz, Poland
Antoni Szydło	Wrocław University of Technology, Poland
Grzegorz Ślaski	Poznań University of Technology, Poland
Paweł Śniady	Wrocław University of Environmental and Life Sciences, Poland
Andrew P. Tarko	Purdue University, West Lafayette, USA
Frane Urem	Polytechnic of Šibenik, Croatia
Hua-lan Wang	School of Traffic and Transportation, Lanzhou Jiaotong University, Lanzhou, China
Mariusz Wasiak	Warsaw University of Technology, Poland
Adam Weintrit	Gdynia Maritime University, Poland
Andrzej Więckowski	AGH University of Science and Technology, Cracow, Poland
Katarzyna Wegrzyn-Wolska	Engineering School of Digital Science Villejuif, France
Adam Wolski	Polish Naval Academy, Gdynia, Poland
Olgierd Wyszomirski	University of Gdańsk, Poland
Elżbieta Załoga	University of Szczecin, Poland
Stanisława Zamkowska	Kazimierz Pulaski University of Technology and Humanities in Radom, Poland
Jacek Żak	Poznań University of Technology, Poland
Jolanta Żak	Warsaw University of Technology, Poland

Referees

Rahmi Akçelik
Przemysław Borkowski
Tony Castillo-Calzadilla
Ireneusz Celiński
Jacek Chmielewski
Piotr Czech
Magdalena Dobiszewska
Michal Fabian
Barbara Galińska
Anna Granà
Róbert Grega
Mehmet Serdar Güzel
Katarzyna Hebel
Peter Jenček
Nan Kang
Peter Kaššay
Jozef Kuĺka
Michał Maciejewski
Elżbieta Macioszek

Krzysztof Małecki
Martin Mantič
Paola di Mascio
Silvia Medvecká-Beňová
Maria Mrówczyńska
Vitalii Naumov
Katarzyna Nosal Hoy
Ander Pijoan
Hrvoje Pilko
Michal Puškár
Alžbeta Sapietová
Grzegorz Sierpiński
Izabela Skrzypczak
Marcin Staniek
Dariusz Tłoczyński
Andrzej Więckowski
Grzegorz Wojnar
Adam Wolski
Ninoslav Zuber

Contents

Traffic Engineering as the Source of Decision in Transport Planning

Applying Random Parameters Model to Evaluate the Impact of Traffic, Geometric and Pavement Condition Characteristics on Accident Frequencies Occurred at A - Roads Networks in the UK

Hamid Ahmed Awad Alacash[1(⊠)] and Tony Parry[2]

[1] College of Engineering, University of Anbar, Ar-Ramadi, Iraq
hamid.awad@uoanbar.edu.iq
[2] Nottingham Transportation Engineering Centre,
University of Nottingham, Nottingham, UK
tony.parry@nottinghamac.uk

Abstract. This study proposed a fixed and random parameters model to analyze multiperiod dry and wet accidents. Applied random parameters model as a new approach in the UK to estimate accident frequencies provides a reasonable understanding of the main factors that affect accident frequencies. Therefore, this approach allows the research to identify and control for contribution factors that may be bias estimation. To demonstrate the application of the proposed models, a six-year (2005–2010) of accidents data from traffic accidents on Norfolk, Oxfordshire and Nottinghamshire A - road network has been used. The estimation results show that random parameters model performs better than fixed parameters models in terms of Akaike information criterion, log-likelihood test, and chi square test. According to parameter estimates the logarithm of the annual average daily flow, percentage of heavy vehicles, gradients and rut depth are positively associated with more dry and wet accidents. On the other hand, speed limit, radius of curvature, number of lanes, skid resistance and texture depth are found as the significant variables decreasing the numbers of dry and wet accidents.

Keywords: Traffic safety · Fixed and random parameters models · Dry and wet accidents

1 Introduction

The frequency of accidents has been studied using different methods by different authors. It is important to note that accident frequency in this paper refers to the number of accidents occurring on roadway segments over a certain period. Most commonly used methods include the Poisson method and negative binomial models. These methods have been used for the prediction of the occurrence of accidents using count data modelling for highways [1–6]. In this study, the random parameters regression model is of great interest given the purpose of this study. Recent studies which have applied the random parameters regression models as their methodologies [7–11].

© Springer Nature Switzerland AG 2020
E. Macioszek and G. Sierpiński (Eds.): Modern Traffic Engineering in the System Approach
to the Development of Traffic Networks, AISC 1083, pp. 3–19, 2020.
https://doi.org/10.1007/978-3-030-34069-8_1

The random parameters models allow the various factors that vary across accidents to be heterogeneous by allowing such parameters to be random. This is in direct constant that parameters are usually constant across the observed road accidents. A demonstration of the random-parameter negative binomial regression as a method for studying the way various factors affect the frequency of accidents on roads is provided in this paper. This paper emphases the application of random parameter modelling as a new approach, estimating accident frequencies in the UK due to the ability of this approach to account and correct for heterogeneity as a result of several factors relating to the characteristics of traffic, geometric and pavement characteristics and condition of the pavement, geometrics, and road traffic.

2   Method

The random parameters models allow the various factors that vary across accident observations to be heterogeneous by allowing such parameters to be random. The method for developing random parameter models is described by Anastasopoulos and Mannering [9]. For the roadway segment i which has n accidents the probability of accidents for the basic Poisson model $P(n_i)$ is:

$$P(n_i) = \frac{EXP(-\lambda_i)\lambda_i^{n_i}}{n_i!} \tag{1}$$

where:
$P(n_i)$ - the Poisson probability of roadway segment i having n accidents,
λ_i - the mean number of accidents in roadway segment i,
n_i - the number of the observed traffic accidents in each roadway segment.

When using a Poisson model, several assumptions must be satisfied. The first assumption is that the network crash data is Equi-dispersed, where λ_i equals the conditional variance. In order for the crash data to be Equi-dispersed, for each roadway segment, the independent variables traffic, geometric and pavement condition characteristics Xj should account for all variations in λ_i. In Eq. 2, the relationship between the λ_i and the Xj can be expressed using a log-linear model, where the natural log of λ_i is linearly related to the product of Xj and the regression coefficients, β_j. The predicted number of traffic accidents at a roadway segment is related to a range of explanatory variables (in this study, traffic, geometric and pavement surface condition characteristics) using the Poisson distribution, typically achieved by using a log-linear function:

$$\lambda_i = EXP(\beta_j X_j) \tag{2}$$

where:
X_j - independent variables,
β_j - estimated regression parameters for the independent variables.

However, a Poisson distribution does not describe traffic accident data because it assumes the data is Equi-dispersed, where the mean of accidents $E(n_i)$ is equal to the variance $VAR(n_i)$ $(E(n_i) = VAR(n_i))$. Traffic accident data is typically over dispersed $(E(n_i) < VAR(n_i))$ and a negative binomial distribution is used to model the data by adding a gamma distributed term to the Poisson model, and the negative binomial model to account for the number of accidents at roadway segments when all independent variables have a fixed effect across the roadway segments is derived by rewriting:

$$\lambda_i = EXP\big(\beta_j X_j + \varepsilon_i\big) \tag{3}$$

where:

ε_i - random error (gamma distributed) term that has mean of 1 and variance σ^2.

In this study, the random parameters modelling technique is used, which allows unobserved heterogeneities to be accounted [12]. A simulated maximum likelihood estimation method was used to incorporate random parameters which may vary across observations in negative binomial and Poisson models. To allow for such random parameters in count-data models, independent parameters can be written as:

$$\beta_{ij} = \beta_j + \varphi_i \tag{4}$$

where:

φ_i - a randomly distributed term (for example a normally distributed term with mean 0 and variance σ^2) [11],

β_{ij} - the mean estimated regression parameters for the independent variables β_j

With this equation, the Poisson parameter becomes $\lambda i/\varphi i = EXP(\beta_j X_j)$ in the Poisson model and $\lambda i/\varphi i = EXP(\beta_j X_j + \varepsilon_i)$ in the negative binomial model with the corresponding probabilities for Poisson or negative binomial now $P(ni/\varphi i)$. With this random-parameter version, the log-likelihood can be written as:

$$LL = \sum_{\forall i} \ln \int_{\varphi_i} g(\varphi_i) P(n_i/\varphi_i) d\varphi_i \tag{5}$$

where:

$g(\varphi_i)$ - the function for the probability density.

Because maximum likelihood estimation of the random parameter Poisson and Negative binomial models' probability estimations are computationally cumbersome, a simulated maximum likelihood method using Halton draws was employed due to the efficient distribution of draws for numerical integration for the statistical models.

The elasticity effect refers to the percentage change in the average number of traffic accidents due to 1 or 10 or 100% change in the independent variable [1]. This term can be calculated using Eq. 6:

$$E_{X_{jk}}^{\lambda_i} = \frac{\partial X_{ji}}{\lambda_i} \cdot \frac{X_{jk}}{\partial X_{jk}} \tag{6}$$

which means the elasticity of accident frequency with respect to the k-th observation for j independent variable for roadway segment i.

3   Data Description

The data in this research consisted of panel data with six years (2005–2010) of annual accidents counts that occurred at roadway segments in the Norfolk, Oxfordshire and Nottinghamshire Counties in the United Kingdom. The data also include traffic, geometric and pavement condition characteristics. For each observation, a total of 10 possible explanatory variables were considered. Traffic accident data from the A - road network in the UK county of Norfolk were collected over a six-year period (2005–2010) from the STATS 19 database (Road traffic accidents datasets-data.gov. uk). Traffic characteristics including annual average daily flow for major and minor direction in the roadway segment, percentage of heavy vehicles and speed limit were obtained from Department of Transport in the UK. Geometric characteristics include radius of curvature, gradient, number of lanes and number of minor access. Finally, the pavement condition characteristics which including skid resistance in term of SCRIM value, rut depth and texture depth. Both geometric characteristics (radius of curvature and gradient) and pavement condition characteristics were obtained from Norfolk, Oxfordshire and Nottinghamshire Councils. A sample summary statistic of variables is presented in Table 1.

Table 1. Summary statistics of the accident, traffic, geometry and pavement condition data for Norfolk, Oxfordshire and Nottinghamshire roadway segments.

County	Variable	Min	Max	Mean	S.D.
Norfolk	Dry accidents	0	15	0.21	0.70
	Wet accidents	0	7	0.10	0.39
	Ln AADF	7.07	9.72	8.40	0.56
	Heavy vehicles %	1	19	5.77	3.89
	Speed limit	32	112	81.49	19.66
	Radius of curvature	46	2000	1096.49	703.7
	Gradient	0	11.2	1.62	1.35
	Number of lanes	1	6	2.09	0.41
	Number of minor access	0	4	0.21	0.48
	Skid resistance (SCRIM value)	0.08	0.62	0.40	0.05
	Rut depth	0.54	90.12	6.06	2.49
	Texture depth	0.13	1.44	0.56	0.21

(*continued*)

Table 1. (*continued*)

County	Variable	Min	Max	Mean	S.D.
Oxfordshire	Dry accidents	0	22	0.26	1.00
	Wet accidents	0	13	0.21	0.73
	Ln AADF	7.12	10.89	8.52	0.63
	Heavy vehicles %	1	11	4.48	2.24
	Speed limit	32	112	81.51	20.45
	Radius of curvature	23	2000	1449.50	1464.66
	Gradient	0.1	13.76	2.38	1.93
	Number of lanes	1	5	2.24	0.58
	Number of minor access	0	7	0.17	0.49
	Skid resistance (SCRIM value)	0.27	0.57	0.41	0.04
	Rut depth	1.2	18.38	8.81	1.94
	Texture depth	0.11	1.30	0.48	0.20
Nottinghamshire	Dry accidents	0	29	0.45	1.55
	Wet accidents	0	17	0.24	0.82
	Ln AADF	7.23	10.77	8.85	0.70
	Heavy vehicles %	1.07	23.50	6.66	4.82
	Speed limit	32	112	82.88	19.36
	Radius of curvature	35	2000	1191.45	856.17
	Gradient	0	11.9	2.02	1.69
	Number of lanes	1	4	2.67	0.97
	Number of minor access	0	5	0.30	0.63
	Skid resistance (SCRIM value)	0.08	0.62	0.40	0.04
	Rut depth	0.58	16.2	6.17	2.12
	Texture depth	0.12	1.36	0.51	0.19

4 Models Results

Table 2 presents the results of fixed parameters models estimation, while the estimation results for random parameters models illustrate in Table 3 for road condition accidents at Norfolk, Oxfordshire and Nottinghamshire.

Table 2. Model estimation results for fixed parameters models for road condition accidents.

Accidents	Variables	Norfolk		Oxfordshire		Nottinghamshire	
		Coefficient	t-value	Coefficient	t-value	Coefficient	t-value
Dry	Constant	−8.682	−20.84	−5.835	−11.31	−4.546	−11.80
	Ln AADF	1.001	20.82	0.571	13.15	0.711	20.03
	Heavy vehicles %	0.009	1.22	0.037	3.02	0.087	18.80
	Speed limit	−0.013	−10.46	−0.006	−4.67	−0.014	−12.51
	Radius of curvature	$-0.6 * 10^{-4}$	−16.16	$-0.1 * 10^{-4}$	−1.20	−0.001	−4.05
	Gradient	0.051	2.84	0.009	0.57	0.017	1.31

(*continued*)

Table 2. (*continued*)

Accidents	Variables	Norfolk		Oxfordshire		Nottinghamshire	
		Coefficient	*t*-value	Coefficient	*t*-value	Coefficient	*t*-value
	Number of lanes	−0.209	−5.11	−0.261	−5.78	−0.115	−4.95
	Number of minor access	0.272	6.39	1.161	20.02	0.033	0.95
	Skid resistance	−1.654	−4.08	−2.815	−3.24	−3.338	−7.53
	Rut depth	0.011	1.13	0.029	1.88	0.001	0.06
	Texture depth	−0.041	−0.33	−0.047	−0.33	−0.490	−3.79
Wet	Constant	−0.900	−15.25	−3.095	−5.36	−3.950	−8.54
	Ln AADF	0.963	14.89	0.223	−2.41	0.558	13.33
	Heavy vehicles %	0.005	0.49	0.043	2.65	0.098	14.08
	Speed limit	−0.012	−6.95	−0.001	−0.79	−0.010	−6.89
	Radius of curvature	−0.001	−9.97	−0.003	−1.60	−0.002	−3.05
	Gradient	0.026	1.10	0.006	-0.04	0.034	2.19
	Number of lanes	−0.146	2.53	−0.607	−12.48	−0.087	−3.00
	Number of minor access	0.144	2.38	0.654	9.41	0.039	0.96
	Skid resistance	−2.188	−3.82	−1.723	−1.99	−3.387	−6.59
	Rut depth	0.003	0.17	0.033	2.07	0.009	0.01
	Texture depth	−0.347	2.38	−0.604	−3.81	−0.544	−3.64

Table 3. Model estimation results for random parameters models for road condition accidents.

Accidents	Variables	Norfolk		Oxfordshire		Nottinghamshire	
		Coefficient	*t*-value	Coefficient	*t*-value	Coefficient	*t*-value
Dry	Constant	−8.764	−21.79	−5.263	−11.31	−3.479	−8.91
	Ln AADF	1.055	23.84	0.581	14.04	0.671	18.23
	Standard deviation	–	–	–	–	–	–
	Heavy vehicles %	0.024	3.00	0.067	5.89	0.132	21.41
	Standard deviation	0.049	13.15	0.058	11.19	1.447	19.72
	Speed limit	−0.021	−16.12	−0.013	−10.45	−0.014	−12.16
	Standard deviation	0.010	33.09	0.010	30.45	–	–
	Radius of curvature	$-0.7*10^{-3}$	−19.12	$-0.2*10^{-4}$	−0.12	$-0.1*10^{-3}$	−5.33
	Standard deviation	$0.4*10^{-3}$	33.09	$0.7*10^{-4}$	5.48	–	–
	Gradient	0.039	2.38	0.001	0.07	0.013	2.06
	Standard deviation	0.041	4.39	0.003	0.38	–	–
	Number of lanes	−0.125	−3.44	−0.120	−3.04	−0.288	−3.88
	Standard deviation	0.115	11.95	0.286	26.72	–	–
	Number of minor access	0.252	6.66	1.123	34.01	0.443	8.31
	Standard deviation	–	–	–	–	–	–
	Skid resistance	−1.386	−3.66	−2.444	−3.20	−4.156	−14.57
	Standard deviation	–	–	–	–	2.605	43.86
	Rut depth	0.008	1.00	0.004	0.25	$0.4*10^{-3}$	−0.04
	Standard deviation	0.015	4.60	0.048	17.38	0.006	1.91
	Texture depth	−0.056	−0.47	−0.788	−5.71	−0.477	−3.89
	Standard deviation	0.036	2.99	1.319	26.45	0.279	6.98

(*continued*)

Table 3. (*continued*)

Accidents	Variables	Norfolk		Oxfordshire		Nottinghamshire	
		Coefficient	t-value	Coefficient	t-value	Coefficient	t-value
Wet	Constant	−8.823	−15.68	−3.093	−6.84	−3.567	−7.92
	Ln AADF	0.990	16.40	0.017	0.43	0.591	13.99
	Standard deviation	–	–	–	–	–	–
	Heavy vehicles %	0.009	0.87	0.038	3.05	0.102	14.34
	Standard deviation	0.030	6.00	0.025	4.91	0.197	4.77
	Speed limit	−0.017	−10.04	−0.010	−7.66	−0.015	−10.82
	Standard deviation	0.009	23.53	0.011	33.92	0.008	24.48
	Radius of curvature	$-0.7*10^{-3}$	−13.39	$-0.7*10^{-4}$	−3.44	$-0.1*10^{-3}$	−4.53
	Standard deviation	$0.5*10^{-3}$	16.16	$0.1*10^{-3}$	10.99	–	–
	Gradient	0.021	0.97	0.006	0.42	0.033	2.28
Wet	Standard deviation	0.053	4.19	0.030	3.81	–	–
	Number of lanes	−0.119	−2.41	−0.651	−19.19	−0.268	−2.52
	Standard deviation	–	–	–	–	–	–
	Number of minor access	0.147	2.77	0.696	20.81	0.306	6.52
	Standard deviation	–	–	–	–	–	–
	Skid resistance	−2.053	−3.99	−1.742	−2.42	−2.645	-11.37
	Standard deviation	–	–	–	–	2.278	32.78
	Rut depth	0.005	0.34	0.041	3.18	0.001	0.07
	Standard deviation	0.019	4.04	–	–	0.008	2.25
	Texture depth	−0.170	−1.10	−0.016	−0.12	−0.091	−3.32
	Standard deviation	0.889	18.08	0.987	21.12	–	–

Table 4 shows summary of performance results for road condition accidents (dry and wet) at Norfolk, Oxfordshire and Nottinghamshire networks. Based on the results that obtained for the overall log-likelihood for random (RPM) and fixed (FPM) parameters models, chi squared, and the confidence level, it's obvious that the random parameters model was preferred to represent the accidents data rather than fixed parameters model.

Table 4. Model estimation summaries.

County	Accident types	Log-L at (β) for FPM	Log-L at (β) for RPM	χ^2	DOF	% Conf
Norfolk	Dry	−7087.64	−7030.99	113.3	7	100
	Wet	−4375.27	−4353.07	44.4	4	100
Oxfordshire	Dry	−6175.67	−6103.88	143.6	7	100
	Wet	−5843.39	−5782.99	120.8	4	100
Nottinghamshire	Dry	−10240.17	−10128.27	223.8	7	100
	Wet	−7340.99	−7306.416	69.1	3	100

A number of variables related to the traffic characteristics (Ln AADF, heavy vehicle percentages and speed limit) were evaluated. These variables were found significant in both dry and wet models for the three counties. The results that obtained from Table 4, revealed that Ln AADF was found to be as a significant fixed parameter at 99% level of confidence for both dry and wet models in the three counties. The results illustrate in Table 4, indicate that there was a positive correlation between Ln AADF and the numbers of dry and wet accidents. These results are in agreement with those obtained by Greibe [13] and Ma et al. [14]. The average elasticity effect of Ln AADF on dry and wet accidents for the three counties are shown in Fig. 1. It can be seen from the figure below that the greatest increase in the number of dry and wet accidents by 8.86 and 8.31% respectively, due to a 1% increase in Ln AADF were showed at Norfolk County.

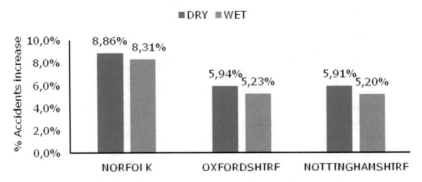

Fig. 1. Estimated elasticity effect of 1% increase in Ln AADF across three networks by road condition accidents

The percentage of heavy vehicle results as random parameters for all dry and wet accident models. For example, the model parameter of heavy vehicle for Norfolk dry accident model was found to vary across roadway segments with a mean of 0.024 and standard deviation of 0.049 (see Table 3). The distribution of this parameter described in Fig. 2 shows that a 31% (positive effect) of the distribution is less than 0 and 69% is greater than 0. This implies that only one third of dry accidents observations for Norfolk roadway segments can be affected positively by the increasing of heavy vehicles percentage. On the other hand, for two third of dry accident observations no positive effect on accidents was noticed due to increasing the heavy vehicles. These results suggest that increasing of heavy vehicles in traffic volume have a partially effect on dry accidents, a possible explanation for this might be that drivers reduced their speed and be more careful when they drive beside the heavy vehicles. These results are consistent with those of Hauer et al. [15], who indicated that the increasing of heavy vehicles is associated with fewer accidents. According to the normal distribution estimation the positive effect of heavy vehicles percentage on dry accident frequencies are 13 and 47% for Oxfordshire and Nottinghamshire respectively. While for wet accidents the positive effects of heavy vehicles are 38, 7 and 30% for Norfolk, Oxfordshire and Nottinghamshire, respectively.

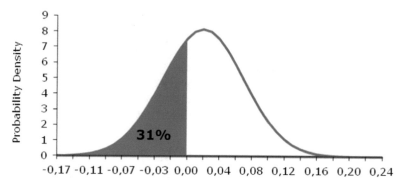

Fig. 2. Distribution of parameter estimation for heavy vehicle percentages of Norfolk dry accidents model

Figure 3 compares the average elasticity effect of heavy vehicle percentage on dry and wet accidents at the three counties. The result shown in Fig. 3 suggests that for both dry and wet accidents, the most effective of a 10% increase in heavy vehicle was at Nottinghamshire County, while the less effective obtained at Norfolk County.

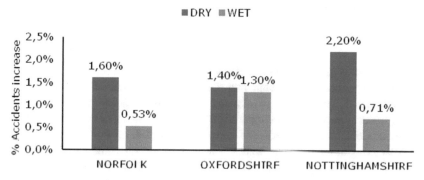

Fig. 3. Estimated elasticity effect of 10% increase in heavy vehicle percentages across three networks by road condition accidents

Roadway segments with lower speed limit are associated with higher numbers of both dry and wet accidents on the three counties. The speed limits parameter resulted as a negative random parameter for all dry and wet models, except for the dry accidents model for Nottinghamshire, where the speed limit parameter resulted as fixed parameter. According to the normal distribution assumption, only 2% is greater than 0, and 98% is less than 0. This implies that, the majority of Norfolk dry accident observations occurs at roadway segments with low speed limit. These results seem to be consistent with other research which found that roadway segments with lower speed conditions are associated with higher accidents [16, 17]. Roadway segments with lower speed limit are found to be positively associated with dry accidents in 10 and 0% for

Oxfordshire and Nottinghamshire roadway segments respectively. In similar context, for wet accidents roadway segments with lower speed limit are found to be positively associated in 2, 1 and 0%, of Norfolk, Oxfordshire and Nottinghamshire roadway segments, respectively. According to the average elasticity effect of speed limit parameter estimates, a 10% increase in the number of roadway segments with lower speed limit increased dry accidents by 17.11, 10.6 and 11.6%; and for wet accidents by 13.85, 8.15 and 12.43%, for Norfolk, Oxfordshire and Nottinghamshire, respectively (see Fig. 4).

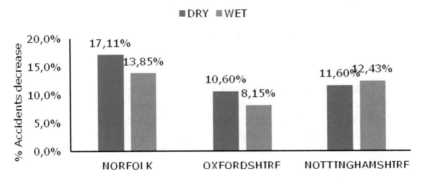

Fig. 4. Estimated elasticity effect of 10% increase in speed limit across three networks by road condition accidents

Negative coefficient on both dry and wet accident models corresponding to radius of curvature for roadway segment indicate that with higher radius of curvature are associated with lower number of dry and wet accidents across the three networks. Radius of curvature resulted as random parameters in all models except in Nottinghamshire models. Increasing radius of curvature at Norfolk network roadway segments are associated with increase dry accidents at 96% of the segments. On the other hand, only 4% of roadway segments this variable was found to be positively associated with dry accidents for Norfolk network. The positive effect for radius of curvature of dry accidents for Oxfordshire and Nottinghamshire are 51 and 0%, and for wet accidents are 8, 32 and 0% for Norfolk, Oxfordshire and Nottinghamshire respectively. These results are in agreement with those obtained by Caliendo et al. [18]. Drivers that travelling on roadway segments with small radius of curvature loss their control on vehicles due to the lateral acceleration which leads to increase the accidents in these locations [19]. As shown in Fig. 5, for dry accidents a decrease of 8.1% on an average for Norfolk network; the decrease is much smaller of 0.03 and 1.6% for Oxfordshire and Nottinghamshire, respectively due to a 10% increase in radius of curvature. A 10% increase in radius of curvature was corresponded to decrease wet accidents of 7.4, 0.9 and 1.6% for Norfolk, Oxfordshire and Nottinghamshire, respectively.

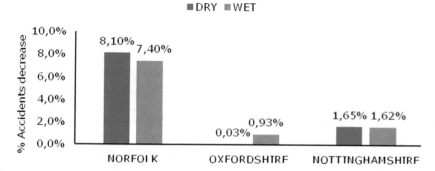

Fig. 5. Estimated elasticity of 10% increase in radius of curvature across three networks by road condition accidents

The significant positive coefficients for gradient indicate that a high value of gradients at roadway segments are associated with higher numbers of dry and wet accidents for the three counties. The effect of gradient was found to be a significant random parameter normally distributed in all models except for Nottinghamshire models, the gradient resulted as fixed parameter. Regarding the Norfolk network a 17% is less than 0 and 83% is greater than 0. These estimated parameters indicated that on only a 17% of roadway segments the increasing of gradients were associated with increasing of dry accidents.

On the other hand, roadway segments with lower gradients were found to be negatively associated with dry accidents at Norfolk network. Similarly, a 37 and 100% are the positive effect of gradients on dry accidents for Oxfordshire and Nottinghamshire, respectively, while for wet accidents a 34, 42 and 100% are found to be the positive effect for Norfolk, Oxfordshire and Nottinghamshire, respectively. As can be seen from Fig. 6 the elasticity effects show that a 10% increase in gradient on Norfolk, Oxfordshire and Nottinghamshire roadway segments resulted in an average increase for dry accidents of 0.62, 0.02 and 0.28%, and for wet accidents of 0.43, 0.14 and 0.67%, respectively.

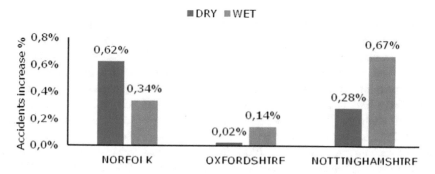

Fig. 6. Estimated elasticity of 10% increase in gradients across three networks by road condition accidents

Regarding to the number of lanes, as can be seen from the random parameters' models, a negative correlation was found both dry and wet accidents and number of lanes. The negative coefficient for the number of lanes width in all the models (see Table 3) indicates that roadway segments more lanes are generally associated with lower crashes. These results are consistent with previous studies that obtained by Ma and Kockelman [20] found that increasing the number of lanes leads to decreases crash counts. In addition, Park et al. [21] found that 4-lane freeways are higher crash occurrence in comparison with 6 lanes. This relationship may be explained due to the fact that in wider roads allow the drivers to make more maneuvers and avoid the conflict between the vehicles. Number of lanes resulted as fixed parameter for all models except for dry models for Norfolk and Oxfordshire where its resulted as random parameters. The distribution of the number of lanes parameter estimates for Norfolk dry accidents model provided more insight. A mean of −0.125 and standard deviation of 0.115, these parameters given that 86% of dry accident observations were negatively associated with number of lanes. On the other hand, 15% of observations were positively associated with number of lanes. The result shown in Fig. 7 suggests that the most effective reduction in dry and wet accidents due to a 100% increase in number of lanes was observed at Nottinghamshire by 76.9 and 71.56%, respectively.

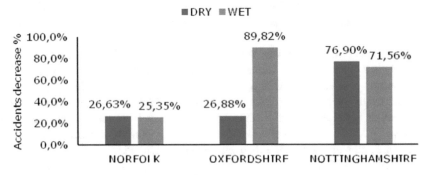

Fig. 7. Estimated elasticity of 100% increase in number of lanes across three networks by road condition accidents

Based on the results of random parameters models illustrate in Table 3, number of minor access in each roadway segments were found to be a highly statistically significant (*t*-value is significant at 99% confidence level) and positively correlated with both dry and wet accident models. These results indicated that the increase in the number of minor access increases dry and wet accident frequencies for roadway segment. These finding are consistent with data obtained in Ackaah and Salifu [22] and Baek and Hummer [23] concluded that higher number of minor access lead to more crashes. Figure 8 provides an overview for the elasticity effect of minor access on dry and wet accident frequencies. A 100% increase in the number of minor access was found to increase the dry accidents by 5.29, 19.09 and 13.73%; and for wet accidents by 3.09, 11.83 and 9.49%, for Norfolk, Oxfordshire and Nottinghamshire network, respectively.

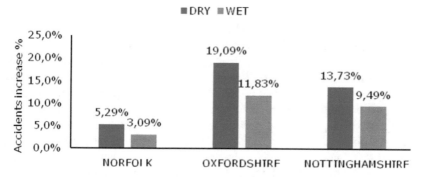

Fig. 8. Estimated elasticity of 10% increase in number of minor access across three networks by road condition accidents

Regarding to the effect of skid resistance on accident frequencies, roadway segments with high SCRIM value were found to have significantly a smaller number of dry and wet accidents in the three counties. As can be seen from Table 3, skid resistance resulted as a fixed parameter for all models except for dry and wet accident models for Nottinghamshire where its resulted as random parameter. For Nottinghamshire network, the skid resistance parameter resulted as random for dry accidents with a mean of -4.156 and standard deviation of 2.605 and for wet accidents with a mean of -2.645 and standard deviation of 2.278. The parameter distribution of indicates that increasing the skid resistance leads to decrease dry accidents at 94% of the roadway segments in Nottinghamshire network. On the other hand, only 6% of roadway segments no decrease in the expected dry accidents were detected due to increasing the skid resistance. Figure 9 compares the average elasticity effect of skid resistance on the expected dry and wet accident for the three counties. The greatest reduction in both dry and wet accidents was observed at Nottinghamshire network by 16.6 and 10.6% respectively, due to a 10% increasing in SCRIM value.

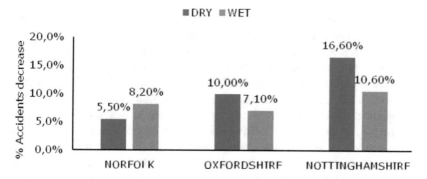

Fig. 9. Estimated elasticity of 10% increase in SCRIM value s across three networks by road condition accidents

The random parameters models result revealed that rut depth was positively correlated with both dry and wet models, this indicated that higher rut depth leads to increase number of accidents. Table 3 illustrates that rut depth resulted as a random parameter for all models, except for Nottinghamshire wet accident models, where it resulted as a fixed parameter. For Norfolk dry accidents model, the mean parameter estimation was 0.008 and standard deviation for the distribution was 0.015. Given these distributional parameters that in about 30% (positive effect) or roadway segments the increasing of rut depth have a positive effect on dry accidents, while in 70% no significant positive effect on dry accidents due to increasing of rut depth. According to the normal distribution parameter estimates, the positive effect for rut depth on dry accidents are 47 and 48% for Oxfordshire and Nottinghamshire, respectively, while for wet accidents are 40 and 45% for Norfolk and Nottinghamshire, respectively. It can be seen from the data in Fig. 10 that for dry accidents the most effective of a 10% increase in rut depth are shown at Norfolk network with 0.5%. For wet accidents, an increase in rut depth by 10% is most effective at Oxfordshire network with 3.4% increase.

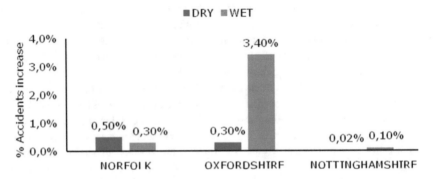

Fig. 10. Estimated elasticity of 10% increase in rut depth across three networks by road condition accidents

Results from both dry and wet models (see Table 3) show that texture depth was found to have significantly decrease accidents. This variable resulted as a random parameter for all models with the exception of wet accidents model where the texture depth resulted as fixed parameter. The modeling results for dry accidents in Norfolk network illustrate that texture depth resulted in with a mean of −0.056 and a standard deviation of 0.036, according to the normal distribution assumption, increasing the texture depth in about 94% of roadway segments leads to decrease dry accidents. On the other hand, on only a 6% (positive effective) of roadway segments there was no detected decrease in dry accidents due to increase the texture depth. The modeling results for dry accidents illustrate that texture depth resulted in with a mean of −0.056, −0.788 and −0.477; and a standard deviation of 0.036, 1.319 and 0.279 for Norfolk, Oxfordshire and Nottinghamshire, respectively. These distribution parameters indicated that a 17 and 4% of roadway segments for Oxfordshire and Nottinghamshire, respectively have a positive effect on dry accidents. For wet accident models, texture depth has a mean of −0.170 and −0.016; and a standard deviation of 0.889 and 0.987 for Norfolk and Oxfordshire, respectively. This indicate that 42 and 49% of roadway segments for Norfolk and Oxfordshire, respectively have a positive effect on wet accidents. Figure 11 shows the average elasticity effect of texture depth on dry and wet accident in the three counties. A 10% increase in texture depth leads to decrease dry accidents of 0.3, 3.8 and 2.4%; and for wet accidents of 1, 0.1 and 0.5%, for Norfolk, Oxfordshire and Nottinghamshire, respectively.

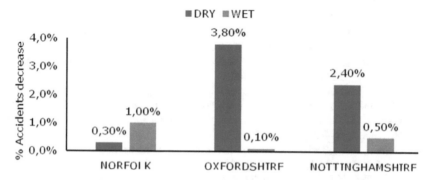

Fig. 11. Estimated elasticity of 10% increase in texture depth across three networks by road condition accidents

5 Conclusions

This study explores the use of random parameters model to examine the traffic, geometric and pavement condition characteristics that significantly influence dry and wet accident frequencies. Using 6 years of accidents data from A-roads networks in Norfolk, Oxfordshire and Nottinghamshire Counties in the UK. The estimation results show that the random parameters models provide a superior statistical fit in comparison with fixed parameters models. The random parameters models indicate that a variety of characteristics relating to traffic are found to significantly influence both dry and wet accident frequencies including Ln AADF, heavy vehicle percentage and speed limit. In terms of geometric factors and their effect on the accident frequencies, radius of curvature, gradient, number of lanes and number of minor access are all found to be statistically significant. In addition, skid resistance, rut depth and texture depth are also, found to have a significant impact on both dry and wet accidents. The results of all dry and wet models reveal that the regression coefficients were positive, for Ln AADF, heavy vehicles percentages, gradients, number of minor access and rut depth. On the other hand, speed limit, radius of curvature, number of lanes, skid resistance and texture depth are associated negatively with both dry and wet accidents in the three counties. Of these variables, Ln AADF, number of minor access are produced statistically significant fixed parameters in all models. On the other hand, heavy vehicle percentage, speed limit, radius of curvature, number of lanes, skid resistance, rut depth and texture depth are generally resulting as random parameters indicating that their impact on accident frequency varies across roadway segments. The main implementation from this paper is first, identify the hazard section based on the accident frequency that obtained from the random parameters model. Second, determine the safety priority based on the reduction in accident frequencies for hazardous sections due to any increase or decrease in the contribution factors (traffic, geometric and pavement characteristics).

References

1. Shankar, V., Mannering, F., Barfield, W.: Effect of roadway geometrics and environmental factors on rural accident frequencies. Accid. Anal. Prev. 27(3), 371–389 (1995)
2. Poch, M., Mannering, F.: Negative binomial analysis of intersection-accident frequencies. J. Transp. Eng. 122(2), 105–113 (1996)
3. Milton, J., Mannering, F.: The relationship among highway geometrics, traffic related elements and motor-vehicle accident frequencies. Transportation 25(4), 395–413 (1998)
4. Abdel-Aty, M.A., Radwan, E.A.: Modeling traffic accident occurrence and involvement. Accid. Anal. Prev. 5(32), 633–642 (2000)
5. Savolainen, P.T., Tarko, A.P.: Safety impacts at intersections on curved segments. Transp. Res. Rec. 1908, 130–140 (2005)
6. Lord, D., Park, Y.J.: Investigating the effects of the fixed and varying dispersion parameters of Poisson-gamma models on empirical Bayes estimates. Accid. Anal. Prev. 40(4), 1441–1457 (2008)
7. Chen, E., Tarko, A.P.: Modeling safety of highway work zones with random parameters and random effects models. Anal. Meth. Accid. Res. 1, 86–95 (2014)
8. Venkataraman, N., Ulfarsson, G., Shankar, V.: Random parameter models of interstate crash frequencies by severity, number of vehicles involved, collision and location type. Accid. Anal. Prev. 59, 309–318 (2013)
9. Anastasopoulos, P.C., Mannering, F.: A note on modeling vehicle accident frequencies with random parameters count models. Accid. Anal. Prev. 41(1), 153–159 (2009)
10. El-Basyouny, K., Sayed, T.: Accident prediction models with random corridor parameters. Accid. Anal. Prev. 41(5), 1118–1123 (2009)
11. Washington, S.P., Karlaftis, M.G., Mannering, F.: Statistical and Econometric Methods for Transportation Data Analysis. CRC Press, New York (2010)
12. Greene, W.: Limdep, Version 9.0. Econometric Software Inc., Plainview, New York (2007)
13. Greibe, P.: Accident prediction models for urban roads. Accid. Anal. Prev. 35(2), 273–285 (2003)
14. Ma, M., Yan, X., Abdel-Aty, M., Huang, H., Wang, X.: Safety analysis of urban arterials under mixed-traffic patterns in Beijing. Transp. Res. Rec. J. Transp. Res. Board 2193, 105–115 (2010)
15. Hauer, E., Council, F., Mohammedshah, Y.: Safety models for urban four-lane undivided road segments. Transp. Res. Rec. J. Transp. Res. Board 1897, 96–105 (2004)
16. Imprialou, M.I.M., Quddus, M., Pitfield, D.E., Lord, D.: Re-visiting crash-speed relationships: A new perspective in crash modelling. Accid. Anal. Prev. 86, 173–185 (2016)
17. Mohammadi, M.A., Samaranayake, V.A., Bham, G.H.: Crash frequency modeling using negative binomial models: An application of generalized estimating equation to longitudinal data. Anal. Meth. Accid. Res. 2, 52–69 (2014)
18. Caliendo, C., Guida, M., Parisi, A.: A crash-prediction model for multilane roads. Accid. Anal. Prev. 39(4), 657–670 (2007)

19. Peters, S.C., Iagnemma, K.: Stability measurement of high-speed vehicles. Veh. Syst. Dyn. **47**(6), 701–720 (2009)
20. Ma, J., Kockelman, K.M.: Bayesian multivariate poisson regression for models of injury count, by severity. Transp. Res. Rec. J. Transp. Res. Board **1950**, 24–34 (2006)
21. Park, B.J., Fitzpatrick, K., Lord, D.: Evaluating the effects of freeway design elements on safety. Transp. Res. Rec. J. Transp. Res. Board **2195**, 58–69 (2010)
22. Ackaah, W., Salifu, M.: Crash prediction model for two-lane rural highways in the Ashanti region of Ghana. IATSS Res. **35**(1), 34–40 (2011)
23. Baek, J., Hummer, J.: Collision models for multilane highway segments to examine safety of curbs. Transp. Res. Rec. **2083**, 128–136 (2008)

The Effect of Delimitation of the Area on the Assessment of the Density of the Road Network Structure

Piotr Soczówka$^{(\boxtimes)}$, Renata Żochowska, Aleksander Sobota, and Marcin Jacek Kłos

Faculty of Transport, Silesian University of Technology, Katowice, Poland
{piotr.soczowka, renata.zochowska, aleksander.sobota, marcin.j.klos}@polsl.pl

Abstract. Road network is a set of objects that are located in a geographical space. The analysis of a road network should be conducted having regard to the area where it exists. The area that is covered in an analysis is usually vast so it may be useful to split it into smaller units. Such delimitation however may have an effect on the assessment of chosen topological measures of road transport network, such as its density. The aim of the paper is to show how a delimitation of a given area may influence the assessment of road network structure that is located in analyzed area.

Keywords: Geographical space · Road network structure · Density of network

1 Introduction

The analysis of the structure of road network is a complex issue. Due to its heterogeneity, complexity and a large scale it often requires diverse actions, such as delimitation of the area where it is located. Because road network connects real locations in geographical space is should not be analyzed without taking into consideration its surroundings.

Proper delimitation of the area is a necessary step to identify its homogeneous aspects and conduct further analysis in smaller but more homogeneous territorial units. Delimitation of geographical space has been elaborated in detail in Sect. 2 of this paper.

Road transport network may be analyzed in terms of its density. Density of road transport network is a very useful measure that allows to assess its complexity. It could be also applied for studies focused on transport accessibility, especially in a spatial aspect of accessibility. Chosen measures of the density of road network have been presented and discussed in Sect. 3 of this paper.

The main goal of the paper was to present the influence of the method of delimitation of the area on numerical values of chosen indices of the density of the structure of transport network. Basic fields of the same shape but of different sizes were

© Springer Nature Switzerland AG 2020
E. Macioszek and G. Sierpiński (Eds.): Modern Traffic Engineering in the System Approach to the Development of Traffic Networks, AISC 1083, pp. 20–36, 2020.
https://doi.org/10.1007/978-3-030-34069-8_2

analyzed. The paper also contains results of case study that has been carried out in order to determine the effect of delimitation of the area on the assessment of the road transport network in terms of its density.

2 Delimitation of Geographical Space

The geographical space should be understood as a real heterogeneous space, taking into account both the diversity of the natural environment and the human environment. Therefore, it may be distinguished in it [1, 2]:

- the ecological space (natural), where nature's laws dominate,
- the economic space, in which economic activity is carried out.

The most frequently cited features of space include [3–6]: continuity, finiteness, resistance, fulfilment, diversification, structure, accessibility, quality, dynamics (variability) and function (potential, planned and present). All divisions and classifications of geographical space should be preceded by the delimitation of its boundaries, i.e. by the determination of the range of occurrence. It is important to define the criteria for the division, which should specify the features and principles that underlie further classification. In general, two ways of delimitation of geographical space may be distinguished:

- division into irregularly shaped territorial units (vector spatial data models),
- division into territorial units with regular shapes (raster spatial data models).

In the case of division of geographical space into irregularly shaped territorial units, it is usually assumed that areas have a specific level of homogeneity due to the analyzed feature (e.g. land use structure, demographic and social characteristics, traffic volume or urbanization level). The method of division of the analysis area also depends strictly on the aim of the research. For example, delimitation of geographical space performed during the construction of a transport model involves spatial and demographic as well as socio-economic analyses, which should make it possible to distinguish specific transport zones (called Traffic Analysis Zones - TAZ's) of irregular shapes that form the basic structure of territorial units in which the travel demand is generated and absorbed. When dividing the area into smaller parts, the following ways of delimitation should be taken into account [7]:

- administrative delimitation - administrative criteria of delimitation on the levels: national, regional/voivodship, sub regional, metropolitan, agglomeration, local, etc.,
- structural delimitation - criteria of structure of settlement units including among others:
  - spatial distribution of density of inhabitants in households,
  - spatial distribution of density of work places,
  - conception of travel generation objects about homogeneous activities/ motivations,
  - natural and artificial barriers for traffic flows (rivers and other infrastructural water objects, roads of the highest classes and categories, railways, undeveloped areas),

- other detailed criteria about creating borders of transport zones located in urbanized zones along or across streets, etc.
- project delimitation - technical and functional criteria of the implemented project based on several characteristics such as functional, organizational, technological, other (e.g. schedule and specific condition of passenger information systems or public transport management system).

The division of geographical space into territorial units of regular shape requires the imposition of a regular grid (i.e. square, hexagonal, triangular, etc.) on the image or the map. Features that are the subject of analysis are assigned to every elementary cell of the grid. Thus, the basic field should be understood as an elementary spatial unit, to which a given value of the analyzed feature or set of attributes is assigned. This approach has been applied, among others, in the analyses of the *urban sprawl* phenomenon, using Cellular Automata (CA) and multi-agent models, for Chicago and Calgary [8–10].

When dividing the area into the fields of equal size, the size of the basic field should be properly determined, depending to a large extent on the possibility of obtaining data at the appropriate level of detail [11]. The size of the basic field is also important in assessing the structure of the road network.

3   Measures for Assessing the Density of the Structure of the Road Network

The transport network is a set of selected objects physically existing in the studied area, which together form a specific spatial structure [12–14]. In other words, it is a set of links and transport points of a given area, and the spatial structure of the transport network shows the connections between its real elements in the geographical space. It follows that the elements of the transport network are also elements of the land use of the area in which they are located [15]. Transport networks are, therefore, real objects occurring in a given area, resulting from linkages, formed between settlement units as a result of social, economic and natural factors [16].

Transport networks may be decomposed due to different criteria. In urban areas, the road network is one of the most important elements of the communication system, and its structure often determines the efficiency of the entire transport network. There are various types of road network structure, including mesh, tree, hub-and-spoke or linear one. In practice, mixed structures are often found, being a combination of basic structures. Each type of network structure may differ in complexity [17].

Assessment of the road network structure requires the construction of its model, which is a simplified representation of reality [14, 16]. By introducing simplifications, it is possible to reproduce only that part of the object or its features whose representation is necessary due to the purpose of the research [18–22]. Most often the transport network is presented as a flow network - using vertices and arches of the graph [23, 24]. In the case of topological assessment, in which the subject of the analysis is the degree of connections between vertices in the spatial system, information about the flow direction is not required, and thus the links of network may be mapped in the form of the graph edges.

For the purpose of assessing the structure of the road network, the area under study has been divided into regular grid with basic fields of equal sizes [16, 25, 26]. In addition, to investigate the effect of the size of the basic field on the values of the road network structure assessment measures, a set of numbers of ways of dividing the analyzed area (geographical space) into basic fields has been introduced:

$$I = \{1, \ldots, i, \ldots, \bar{I}\} \tag{1}$$

where:
i - number of the individual way of dividing the area,
$\bar{I}$ - number of all analyzed ways of dividing the area (the size of the set I).

The set of the basic fields for the i-th way of dividing the area has been described as:

$$R_i = \{1, \ldots, r_i, \ldots, \bar{R_i}\}, \quad i \in I \tag{2}$$

where:
r_i - number of the individual basic field for the i-th way of dividing the area,
$\bar{R_i}$ - number of all basic fields analyzed for the i-th way of dividing the area (the size of the set R_i).

In each r_i-th basic field, both nodes representing real infrastructure elements (transport vertices) as well as the points of intersection of the linear infrastructure elements with the border of the area (boundary vertices) have been identified. Therefore, the graph of the road network structure for the r_i-th basic field may be written as:

$$G_{r_i} = (W_{r_i}, V_{r_i}, K_{r_i}), \quad r_i \in R_i, \quad i \in I \tag{3}$$

where:
W_{r_i} - set of transport vertices of the graph G_{r_i},,
V_{r_i} - set of boundary vertices of the graph G_{r_i},
K_{r_i} - set of edges of the graph G_{r_i}.

All vertices in each of the basic fields of the graph G_{r_i} have been appropriately numbered. The set of transport vertices of the graph G_{r_i} is shown as:

$$W_{r_i} = \{1, \ldots, w_{r_i}, \ldots, \bar{W_{r_i}}\}, \quad r_i \in R_i, \quad i \in I \tag{4}$$

where:
w_{r_i} - number of the individual transport vertex of the graph G_{r_i},
$\bar{W_{r_i}}$ - number of all transport vertices of the graph G_{r_i} (the size of the set W_{r_i}).

The set of boundary vertices of the graph G_{r_i} is written as:

$$V_{r_i} = \{\bar{W_{r_i}} + 1, \ldots, v_{r_i}, \ldots, \bar{W_{r_i}} + \bar{V_{r_i}}\}, \quad r_i \in R_i, \quad i \in I \tag{5}$$

where:

v_{r_i} - number of the individual boundary vertex of the graph G_{r_i},

$\overline{V}_{r_i}$ - number of all boundary vertices of the graph G_{r_i} (the size of the set V_{r_i}).

In turn, the set of the edges of the graph G_{r_i} is determined as:

$$K_{r_i} = \left\{1, \ldots, k_{r_i}, \ldots, \overline{K_{r_i}}\right\}, \quad r_i \in R_i, \quad i \in I \tag{6}$$

where:

k_{r_i} - number of the individual edge of the graph G_{r_i},

$\overline{K}_{r_i}$ - number of all edges of the graph G_{r_i} (the size of the set $K_{r_i}r$).

The density is one of the most important measures to assess the complexity of structure of the transport network [26–28]. There are different ways to determine it. One of the measures depends on the length of arches (edges) and the surface of the analyzed area and is expressed in [km/km^2] using the formula:

$$NL_{r_i}(i) = \frac{\sum_{k_{r_i}=1}^{\overline{K_{r_i}}} l_{k_{r_i}}}{a_i}, \quad r_i \in R_i, \quad i \in I \tag{7}$$

where:

$\overline{K}_{r_i}$ - number of all edges of the graph G_{r_i} representing the structure of the r_i-th basic field for the i-th way of dividing the area,

$l_{k_{r_i}}$ - the length of the edge k_{r_i} of the graph G_{r_i} representing the structure of the r_i-th basic field for the i-th way of dividing the area [km],

a_i - surface area of the basic field for the i-th way of dividing the area [km^2].

Another measure of network density depends on the number of vertices and the surface of area under study. In this case, the density is expressed in $\left[\frac{1}{km^2}\right]$ using the formula:

$$NV_{r_i}(i) = \frac{\overline{W}_{r_i} + \overline{V}_{r_i}}{a_i}, \quad r_i \in R_i, \quad i \in I \tag{8}$$

where:

$\overline{W}_{r_i}$ - number of all transport vertices of the graph G_{r_i} representing the structure of the r_i-th basic field for the i-th way of dividing the area,

$\overline{V}_{r_i}$ - number of all boundary vertices of the graph G_{r_i} representing the structure of the r_i-th basic field for the i-th way of dividing the area,

a_i - surface area of the basic field for the i-th way of dividing the area [km^2].

Analogously, the density of the transport network may be expressed taking into account the number of edges and surface of the area in which the analyzed edges occur. In this case the unit is $\left[\frac{1}{km^2}\right]$. The following formula is used:

$$NE_{r_i}(i) = \frac{\overline{K_{r_i}}}{a_i}, \quad r_i \in R_i, \quad i \in I \tag{9}$$

where:

$\overline{K_{r_i}}$ - number of all edges of the graph G_{r_i} representing the structure of the r_i-th basic field for the i-th way of dividing the area,

a_i - surface area of the basic field for the i-th way of dividing the area [km^2].

4 Case Study

A study has been carried out in order to determine the influence of the size of basic field on numerical values of topological indices that allow to assess the density of the structure of transport network. The area of study has been delimited in the center of Katowice. It is a square with a side of 2,000 m. That allowed to split the area of study into regular grid with basic fields being squares of four different sizes:

- case 1: basic field with a side of 100 m,
- case 2: basic field with a side of 200 m,
- case 3: basic field with a side of 400 m,
- case 4: basic field with a side of 500 m.

In Fig. 1 the area of study has been presented for each case of analysis and before splitting it into regular units.

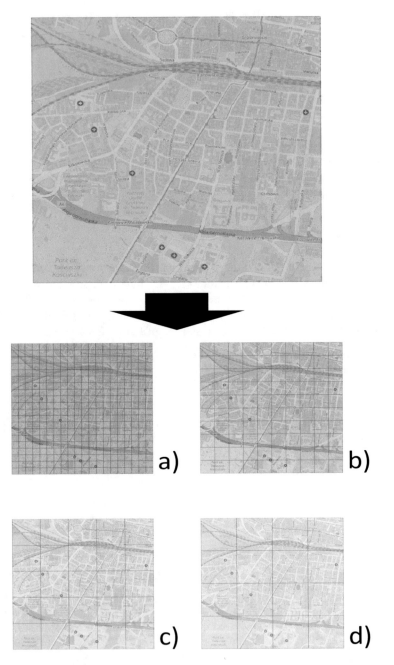

Fig. 1. Area of study in each case of analysis: (a). basic field with a side of 100 m, (b). basic field with a side of 200 m, (c). basic field with a side of 400 m, (d). basic field with a side of 500 m

Basic information about each case of analysis (case 1 - case 4) has been presented in Table 1.

Table 1. Basic information about cases of analysis.

Case	Number of basic fields [-]	Side of basic field [m]	Area of basic field [km²]	Number of transport vertices [-]	Number of boundary vertices [-]	Number of edges [-]
1.	400	100	0.01	253	1,209	1,008
2.	100	200	0.04	253	611	717
3.	25	400	0.16	253	316	578
4.	16	500	0.25	253	307	573

Transport vertices represent real infrastructure element, therefore number of such vertices is constant in each case. Number of boundary vertices, as well as number of edges, however may vary, depending on the number of basic fields. It should be noted that if a number of basic fields decreases also number of boundary vertices and number of edges decreases.

Five levels of density have been assumed and each color denotes different level of density, as presented in Table 2.

Table 2. Levels of density.

Level	Color	NL [km/km²]	NV [1/km²]	NE [1/km²]
1.		<0–10>	<0–300>	<0–250>
2.		(10–20>	(300–600>	(250–500>
3.		(20–30>	(600–900>	(500–750>
4.		(30–40>	(900–1,200>	(750–1,000>
5.		(40–50>	(1,200–1,500>	(1,000–1,250>

Figure 2 presents levels of density calculated using all three measures in each basic field in case 1. According to Fig. 2 in most fields level of density is relatively low, usually levels 1 or 2. It is observed that level of density of transport network in particular field may vary, depending on the measure of density that was used.

Figure 3 presents levels of density in each basic field in case 2.

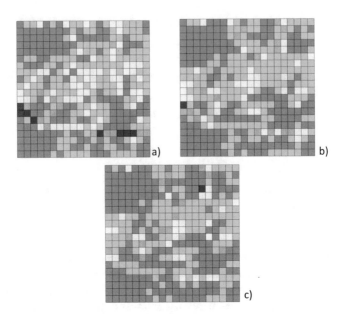

Fig. 2. Levels of density of transport network in each basic field, case 1: (a) measure NL, (b) measure NV, (c) measure NE

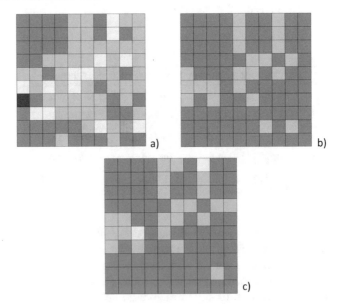

Fig. 3. Levels of density of transport network in each basic field, case 2: (a) measure NL, (b) measure NV, (c) measure NE

When comparing Figs. 2b, c with 3b and c it should be noted that levels of density of transport network are significantly lower in case 2. Values of measure NV have not exceeded level 2 whereas values of measure NE have exceeded level 2 only in two basic fields. Lower values of density are caused by the fact that in case 2 the number of boundary vertices and number of edges is smaller than in case 1 whereas the area is twice bigger than in case 1. Figure 4 presents levels of density of transport network in case 3.

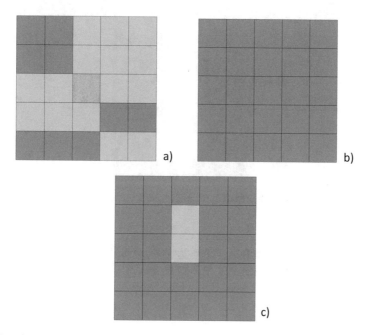

Fig. 4. Levels of density of transport network in each basic field, case 3: (a) measure NL, (b) measure NV, (c) measure NE

Levels of density of transport network calculated using measures NV and NE are very low in case 3. Values of measure NV haven't exceed 300/km^2 in any basic field. It is also associated with the fact that area of basic fields is bigger whereas number of boundary vertices is smaller than in cases 1 or 2.

Levels of density of transport network in case 4 have been presented in Fig. 5.

Values of density of transport network in the area of study, calculated using measures NL, NV and NE in case of analysis 4 are similar to case 3. Values of each measure have not exceeded level 2 in any basic field.

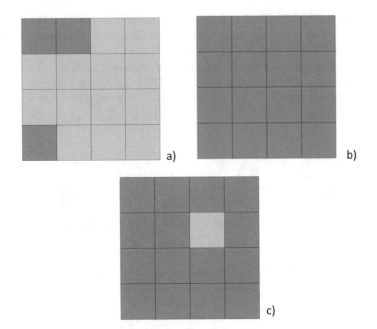

Fig. 5. Levels of density of transport network in each basic field, case 4: (a) measure NL, (b) measure NV, (c) measure NE

For each basic field in cases 1–4 average values of density measures NL, NV and NE have been calculated, according to formulas (10), (11) and (12):

$$NL_{SR}(i) = \frac{\sum_{r_i=1}^{\overline{R}_i} NL_{ri}(i)}{\overline{R}_i}, \quad r_i \in R_i, \quad i \in I \tag{10}$$

$$NV_{SR}(i) = \frac{\sum_{r_i=1}^{\overline{R}_i} NV_{ri}(i)}{\overline{R}_i}, \quad r_i \in R_i, \quad i \in I \tag{11}$$

$$NE_{SR}(i) = \frac{\sum_{r_i=1}^{\overline{R}_i} NE_{ri}(i)}{\overline{R}_i}, \quad r_i \in R_i, \quad i \in I \tag{12}$$

According to Table 3, average NL density was very similar in each case of analysis. That could be explained by the fact that total length of network in each case was the same - only number of vertices and edges could change, but those aspects are not used for calculating NL measure.

It should be noted that NV density decreases if a side of a basic field is increased. To calculate measure NV vertices are taken into account and according to Table 1, the number of boundary vertices decreases in each case.

Table 3. Comparison of density of transport network in each case of analysis.

Case	Average value of NL [km/km^2]	Average value of NV [1/km^2]	Average value of NE [1/km^2]
1.	12.36	365.50	252.00
2.	12.36	216.00	179.50
3.	12.36	142.25	144.50
4.	12.36	140.00	143.50

As it has been presented, the average value of NE measure changes in a very similar way to NV measure. In both cases a significant decrease of average values has been observed.

Table 4 presents comparison of average levels of density in each case of analysis.

Table 4. Comparison of average level of density.

Case	Average level of NL	Average level of NV	Average level of NE
1.	1.89	1.77	1.63
2.	1.79	1.22	1.27
3.	1.68	1.00	1.08
4.	1.81	1.00	1.06

Average levels of density NL and NE decreases as the area of basic fields increases. On the other hand, the average level of density NL in case 4 is higher than in cases 2 and 3.

Each basic field with a side of 100 m is a part of bigger field, with side of 200 m, 400 m or 500 m. Four 100 m basic fields grouped together create 200 m basic field, 16 such fields create one 400 m field, and 25 constitute a 500 m field. For measures of density a study was carried out to learn about differences that appear when 100 m fields are grouped together to create bigger fields.

Three cases have been studied:

- case 1: grouping together 100 m fields to create 200 m fields,
- case 2: grouping together 100 m fields to create 400 m fields,
- case 3: grouping together 100 m fields to create 500 m fields.

For each case figures presenting differences between analyzed fields have been prepared. Each basic field with a side of 100 m has one of three colors:

- red color denotes that a level of density of transport network in a particular 100 m basic field is smaller than in a bigger field (200 m, 400 m or 500 m) that a given 100 m is a part of,
- green color denotes that a level of density of transport network in a particular 100 m basic field is bigger than in a bigger field (200 m, 400 m or 500 m) that a given 100 m is a part of,

- yellow color denotes that a level of density of transport network in a particular 100 m basic field is the same as in a bigger field (200 m, 400 m or 500 m) that a given 100 m is a part of.

Figure 6 presents change of level of density of transport network for 100 m basic fields when they are used to create 200 m basic fields.

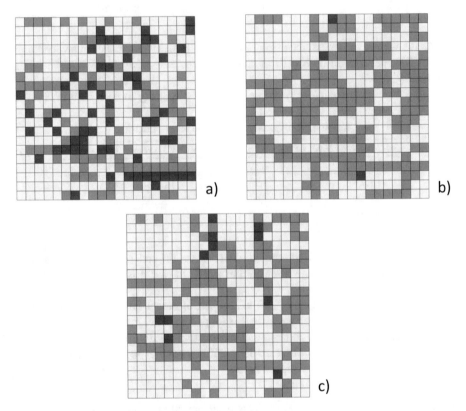

Fig. 6. Levels of density of transport network in each basic field with a side of 100 m, compared to level of density in a basic field with a side of 200 m: (a) measure NL, (b) measure NV, (c) measure NE

In case of NL measure, density of transport network in about 39% of 100 m basic fields is different than in 200 m basic fields. In about 60% of instances when the level density has changed, it is greater in smaller, 100 m field, than in 200 m basic fields that given 100 m fields are a part of. In case of NV density of transport network, change has been observed in 46% of 100 m basic fields. In all but three instances a level of NV density in 100 m basic fields is greater than in 200 m basic fields. The smallest number of differences has been observed for a NE density - density of transport network in only about 37% of 100 m basic fields is different than in 200 m basic fields that are created using 100 m fields. In about 90% instances of such difference, a greater level of density has been calculated for a 100 m basic field.

Figure 7 presents comparison of levels of density of transport network for 100 m basic fields when they are used to create 400 m basic fields.

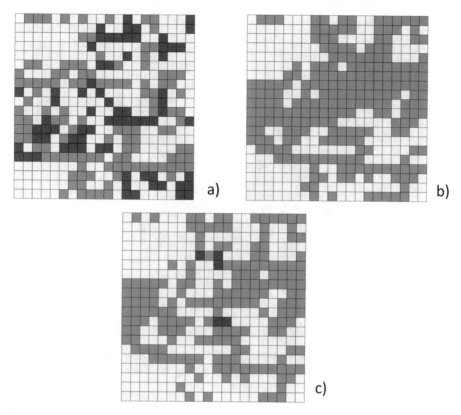

Fig. 7. Levels of density of transport network in each basic field with a side of 100 m, compared to level of density in a basic field with a side of 400 m: (a) measure NL, (b) measure NV, (c) measure NE

NL density of transport network in about 48% of 100 m basic fields is different than in 400 m basic fields. It is a growth of 9% points in comparison to case 1. In about 62% of instances when the level density has changed, it is greater in 100 m field, than in 400 m basic fields that are created using 100 m fields. In case of NV density of transport network, a change has been observed in 56% of 100 m basic fields. It is a difference of 10% points in comparison to case 1. In all instances a level of NV density in 100 m basic fields is greater than in 400 m basic fields. NE density of transport network is different in about 46% of 100 m basic fields than in 400 m basic fields that are created using 100 m fields. In all but five instances of such difference, a greater level of density has been calculated for a 100 m basic field.

Figure 8 presents comparison of levels of density of transport network for 100 m basic fields when they are used to create 500 m basic fields.

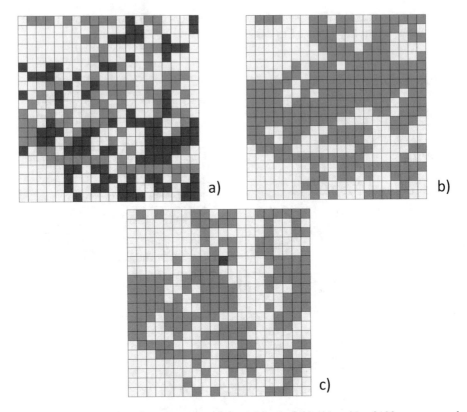

Fig. 8. Levels of density of transport network in each basic field with a side of 100 m, compared to level of density in a basic field with a side of 500 m: (a) measure NL, (b) measure NV, (c) measure NE

In case of NL density of transport network in about 49% of 100 m basic fields is different than in 500 m basic fields. It is a growth of 10 percentage points in comparison to case 1. In about 62% of instances when the level density has changed, it is greater in 100 m field, than in 500 m basic fields that are created using 100 m fields. In case of NV density of transport network, a change has been observed in 56% of 100 m basic fields - exactly like in case 2. In all instances a level of NV density in 100 m basic fields is greater than in 500 m basic fields. NE density of transport network is different in about 44% of 100 m basic fields than in 500 m basic fields that are created using 100 m fields. It is still a bigger percentage share than in case 1, however more differences has been observed for case 2. In all but one instances of such difference, a greater level of density has been calculated for a 100 m basic field.

5 Conclusion

Delimitation of the area for purposes of the analysis of road network is a complex issue. Significant number of possibilities of conducting such delimitation lead to decision problem. The aim of the paper was to learn if delimitation of the area influences topological measures of the road transport network, i.e. its density.

The results presented in the Sect. 4 of the paper prove that such influence does exist. For purposes of the study an area in Katowice was covered with a regular grid of squares with different side to create individual units - basic fields. The results of calculation of three different density measures for each case of analysis have been presented. In each case of analysis the sides of squares were changed. The results of the study show also how the density changes in squares of different sides. It has been shown that the bigger the side of the square the lower the level of density is.

A comparison of levels of density for individual basic fields in each case was also conducted.

Further research should focus on examination of other topological measures, i.e. connectivity to provide a full study of dependences between delimitation of the area and measures for assessment of a road network.

References

1. Bajerowski, T., Kowalczyk, A.M., Ogrodniczak, M.: Network structures in developing uniformed service intervention maps. In: Proceedings of International Multidisciplinary Scientific GeoConference: SGEM: Surveying Geology & Mining Ecology Management, vol. 17, pp. 619–624. SGEM World Science, Sofia (2017)
2. Kuciński, K.: Geografia ekonomiczna. Zarys teoretyczny. Szkoła Główna Handlowa w Warszawie, Warsaw (2004)
3. Borsa, M.: Gospodarka i Polityka Przestrzenna. Część I. Gospodarka przestrzenna. Wyższa Szkoła Społeczno-Ekonomiczna, Warszawa (2004)
4. Small, K.: Urban Transportation Economics. Taylor & Francis, Milton Park (2013)
5. Domański, R.: Gospodarka Przestrzenna. Podstawy teoretyczne. Wydawnictwo Naukowe PWN, Warszawa (2006)
6. Fujita, M., Krugman, P., Venables, A.J.: The Spatial Economy Cities Regions and International Trade. The MIT Press, Cambridge (1999)
7. Karoń, G., Żochowska, R., Sobota, A., Janecki, R.: Selected aspects of the methodology for delimitation of the area of urban agglomeration in transportation models for the evaluation of ITS projects. In: Sierpiński, G. (ed.) Advanced Solutions of Transport Systems for Growing Mobility. AISC, vol. 631, pp. 243–254. Springer International Publishing (2018). https://doi.org/10.1007/978-3-319-62316-0_20
8. Batty, M.: Agents, cells and cities: new representational models for simulating multi-scale urban dynamisc. Environ. Plann. A: Econ. Space 37(8), 1373–1394 (2005)
9. Batty, M., Torrens, P.M.: Modelling Complexity: The Limits to Prediction. CASA Working Paper 36. Centre for Advanced Spatial Analysis. University College London, London (2001)
10. Torrens, P.M.: Simulating sprawl. Ann. Assoc. Am. Geogr. 96(2), 248–275 (2006)
11. Liszewski, S.: Geografia urbanistyczna. Wydawnictwo Naukowe PWN, Warszawa (2012)
12. Steenbrink, P.: Optymalizacja sieci transportowych. WKiŁ, Warszawa (1978)

13. Papageorgiou, G., Damianou, P., Pitsillides, A., Aphamis, T., Charalambous, D., Ioannou, P.: Modelling and simulation of transportation systems: a scenario planning approach. Automatika **50**(1–2), 39–50 (2009)

14. Leszczyński, J.: Modelowanie systemów i procesów transportowych. Oficyna Wydawnicza Politechniki Warszawskiej, Warszawa (1999)

15. Borowska-Stafańska, M., Wiśniewski, S.Z.: The use of network analysis in the process of delimitation as exemplified by the administrative division of Poland. Geodesy and Cartography, **66**(2), 155–170 (2017)

16. Ratajczak, W.: Modelowanie sieci transportowych. Wydawnictwo Naukowe UAM, Poznań (1999)

17. Żochowska, R., Soczówka, P.: Analysis of selected structures of transportation network based on graph measures. Sci. J. Silesian Univ. Technol. Ser. Transp. **98**, 223–233 (2018)

18. Jacyna, M., Wasiak, M., Lewczuk, K., Kłodawski, M.: Simulation model of transport system of Poland as a tool for developing sustainable transport. Arch. Transp. **31**(3), 23–35 (2014)

19. Jacyna-Gołda, I., Izdebski, M., Podviezko, A.: Assessment of efficiency of assignment of vehicles to tasks in supply chains: a case study of a municipal company. Transport **32**, 243–251 (2017)

20. Jachimowski, R., Kłodawski, M.: Simulated annealing algorithm for the multi-level vehicle routing problem. Prace Naukowe Politechniki Warszawskiej, Seria Transport **97**, 195–204 (2013)

21. Jacyna, M., Wasiak, M., Lewczuk, K., Karoń, G.: Noise and environmental pollution from transport: decisive problems in developing ecologically efficient transport systems. J. VibroEng. **19**(7), 5639–5655 (2017)

22. Gołda, P., Manerowski, J.: Support of aircraft taxiing operations on the apron. J. KONES **21**(4), 127–135 (2014)

23. Jacyna-Gołda, I., Żak, J., Gołębiowski, P.: Models of traffic flow distribution for various scenarios of the development of proecological transport system. Arch. Transp. **32**, 17–28 (2014)

24. Jachimowski, R., Żak, J., Pyza, D.: Routes planning problem with heterogeneous suppliers demand. In: 21st IEEE International Conference on Systems Engineering, pp. 434–437. IEEE, USA (2011)

25. Laurini, R., Thompson, D.: Fundamentals of Spatial Information Systems. Academic Press Harcourt Brace & Company Publishers, Cambridge (1998)

26. Gavu, E.: Network based indicators for prioritising the location of a new urban connection: case study Istanbul, Turkey. International Institute for Geo-Information Science and Earth Observation, Enschede (2010)

27. Rodrigue, J.P., Comtois, C., Slack, B.: The Geography of Transport Systems. Routledge, London (2006)

28. Frazila, R.B., Zukhruf, F.: Measuring connectivity for domestic maritime transport network. J. Eastern Asia Soc. Transp. Stud. **11**, 2363–2376 (2015)

Resistance Probabilistic Analysis
of RCC Pavement

Izabela Skrzypczak[1(✉)], Wanda Kokoszka[1], Tomasz Pytlowany[2],
and Wojciech Radwański[2]

[1] Faculty of Civil and Environmental Engineering and Architecture,
Rzeszów University of Technology, Rzeszów, Poland
{izas,wandak}@prz.edu.pl
[2] Polytechnic Institute, Krosno State College, Krosno, Poland
tompyt@pwsz.krosno.pl

Abstract. Pavements are an essential feature of the urban communication system and provide an efficient means of transportation of goods and services. Depending on its rigidity compared to the subsoil, pavements are classified as flexible, rigid and semi-flexible. The Rigid concrete pavements are now a day becoming more popular in Poland because of reliability construction. Concrete roads are called rigid. The largest advantages to using concrete pavement are in its durability and ability to hold a shape. The conventional methods of rigid pavement Analysis and design are based on the closed-form solutions obtained from the static analysis of infinitely long plates resting on an elastic foundation. In recent years computing programs and probability analysis emerged as a tool which is calculation of capturing pavement reliability. The study of the reliability of RCC pavement on using FReET software was presented. Analysis for reliability have been done using probability and semi probability of RCC pavement.

Keywords: Transport model · Travel demand · Transportation zones

1 Introduction

RCC is a rapidly evolving concrete material and construction methodology. Due to its unique nature, it is applied widely in diverse engineering applications ranging from hydraulic to pavement structures. RCC is an economical, fast and durable candidate for many pavement applications. Properly designed RCC mixes can achieve outstanding compressive strengths similar to those of conventional concrete. Due to its relatively coarse surface, RCC has traditionally been used for pavements carrying heavy loads in low-speed areas, such as parking, storage areas, port, airport service areas, intermodal and military facilities [1]. With improved paving and compaction methods as well as surface texturing techniques, recent applications of RCC can be found for interstate highway shoulders, city streets, and rural highways [1]. In addition, due to low water content RCC pavements have reduced shrinkage and low maintenance costs [1, 4]. Growing interest in applications of RCC has lead to many laboratory and investigations studies [1–25]. Moreover this is also motivating exploration of various reliability and durability related studies [2, 3]. This paper is dedicated to an probabilistic analysis

© Springer Nature Switzerland AG 2020
E. Macioszek and G. Sierpiński (Eds.): Modern Traffic Engineering in the System Approach
to the Development of Traffic Networks, AISC 1083, pp. 37–45, 2020.
https://doi.org/10.1007/978-3-030-34069-8_3

concerning with reliability studies. According probabilistically approach, a pavement is safe when the estimated bearing capacity probability is equal to or higher than the load effect level. Thus, the consideration of reliability based design of pavement becomes very important. Estimated reliability value may be justified provided the proper distributions of pavement performance parameters are adopted including their level of confidence or acceptability [26–31]. This paper presents the issues related to probability of failure calculation in RCC pavements. The issues of reliability evaluation that involved with many complexities due to materials, structural and loadings conditions, including uncertainty associated with various input parameters are discussed.

2   Structural Assessment and Reliability Analysis

In the assessment and designing of structures, the probabilistic methods can be used widely. The aim of structural design is to realize structures that meet the expected performance, which can be often represented by a target reliability level [26–35]. There are different approaches for reliability verification: deterministic, semi-probabilistic, and probabilistic.

The most common deterministic safety measure is the global factor of safety, defined as the ratio of the resistance over the load effect. The concept of the allowable stresses is a traditional deterministic method, where failure of the structure is assumed to occur when any stressed part of it reaches the permissible stress. Deterministic verification methods which are based on a single global safety factor do not properly account for the uncertainties associated with strength and load evaluation. The semi-probabilistic approach is based on the limit state principle and makes use of partial safety factors for checking the structural safety. These partial factors have been calibrated so that a structure that satisfies the safety check using a set of design parameters will also satisfy the target reliability level. The semi-probabilistic verification method is still a simplified method but it can much better account for the uncertainties of some design parameters. Probabilistic verification procedures are also based on the principle of limit states, by checking that predefined target structural reliability levels are not exceeded. This approach takes into account explicitly the uncertainties.

Regardless of the uncertainties in different parameters accounting for the analysis and design of a structure, it is very difficult to measure its absolute safety using deterministic analysis. Therefore, one of the most important ways to specify a rational criterion for ensuring the safety of a structure is evaluating its reliability or probability of failure. The reliability of a structure is defined as its ability to fulfill the design purpose for some specified design lifetime [26–35]. Reliability is often understood as the probability of not losing the intended function of the structure.

Most of the modern codes for constructions have recognized the need of using advanced reliability based design methods that allow taking into account various sources of uncertainty. To verify whether or not a structural design is acceptable, the uncertainties are modeled by using statistical tools and the failure probability is estimated with respect to all relevant limit states. The three main documents that have been drawn on reliability based design, are the standard ISO 2394 [35], the probabilistic model code developed by the Joint Committee on Structural Safety (JCSS) [36] and the structural Eurocodes [32, 33].

3   Probabilistic Analysis of RCC Pavement

In pavement design process, a pavement is idealized as multilayered structure and, the fatigue is considered as primary mode of structural failure. Due to large uncertainty associated with variables, the deterministic approach seems to be inadequate. While adopting the probabilistic approach, it needs to consider the probability distribution of particular variables, and its distribution parameters.

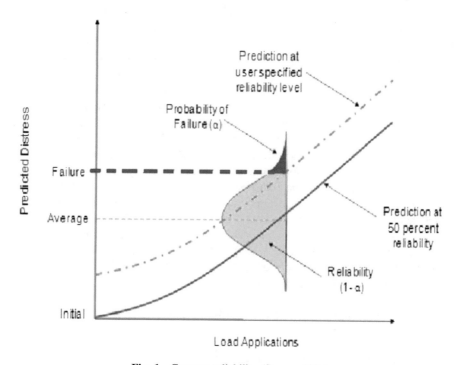

Fig. 1. Concept reliability (Source: [39])

In many publications [37, 38] the distribution of random variables is considered as normally or lognormally distributed. A practical method to consider the uncertainties and variation in design is needed to provide an acceptable level of risk that the pavement will meet its performance expectations. Since the 1970s, reliability and risk analysis concepts have been applied in both empirical and mechanistic-empirical design approaches for design of flexible and rigid pavement structures [37, 38], and more and more computer tools have been developed to facilitate the analysis process. A concrete pavement failure primarily happens due to fatigue and cracking distress. However, the prediction of failure against each mode is very uncertain due to significant inherent variability associated with large number of input parameters [20–38]. Thus, it demands a reliability based design approach, which can accommodate such variability. The motivation of this paper is to investigate the applicability of the analytical method, the Method of Moments, to estimate the failure probability and to model pavement reliability function. The advantage of such an approach is the ability to use the reliability function in RCC pavement. Reliability is defined as the probability that the pavement distress will be less than the critical level of distress over the design period (1):

$$Reliability = Probability \ (Predicted \ Distress < Critical \ Distress) \qquad (1)$$

In Fig. 1 is showed, the mean distress is predicted over the intended pavement design life using the average values (i.e., 50% reliability) for all inputs [39].

Reliability is dependent on the standard error of each distress (including Stress Ratio (SR) or International Roughness Index (IRI)) prediction model as determined from the calibration results. The error in prediction is assumed to be normally distributed for all pavement distresses. The mean value of the distress is increased by the number of standard errors associated with the reliability level [37–39].

Thickness design for RCC pavements may follow the same design strategy as for conventional concrete pavements, i.e., keeping the pavement's flexural stress and fatigue damage caused by wheel loads within an allowable limit [37–39]. By flexural stress is meant the tensile stress at the bottom of a RCC slab under traffic loading. The critical (maximum) flexural stress under wheel load divided by flexural strength of the concrete slab is defined as Stress Ratio (SR). A fatigue curve (so called fatigue model) between different allowable load limits and SRs is needed in the thickness design, which can be determined from laboratory beam fatigue tests. The design thickness is then estimated based on the allowable loads to failure at a certain design SR. Both the Portland Cement Association (PCA) [21] and the U.S. Army of Corps Engineering (USACE) [20] developed the thickness design procedures for RCC industrial pavements [21, 39]. In [39] is used the following fatigue model (2):

$$logN_f = 10.25476 - 11.1872(SR) \ for \ SR > 0.38 \qquad (2)$$

where:

N_f - is the allowable number of load repetitions

Formulas for the fatigue model used for RCC pavement thickness design developed by American Concrete Institute (ACI) are Eqs. (3) and (4) [37, 38]:

$$N_f = (4.2577/(SR - 0.4325))3.268 \ for \ 0.45 < SR < 0.55 \qquad (3)$$

$$logN_f = 11.737 - 12.077(SR) \ for \ SR \geq 0.55 \qquad (4)$$

The traffic with different axle loads under the mixed loading condition can be converted into the standard axle load according to Polish catalog of typical rigid surfaces, for traffic category KR2, equivalent axle loads $N_{f \ code}$ is in interval form 60 000 to 280 000 [7]. To evaluate reliability, a limit state function can be expressed as (5):

$$R = Prob.(N_f - Nf\_catalog < 0) \qquad (5)$$

Knowing the distributions of parameters for N_f and $N_{f\_catalog}$ and their COV values, the reliability can be estimated for any given set of input.

4 Case Study of the Reliability of RCC Pavement

The presence of uncertainties in various parameters accounting for the analysis and design, it is very difficult to measure safety for a structure from deterministic analysis. The main objective of this research is to evaluate the safety in terms of reliability of RCC pavement design using [39]. A RCC pavement as a structural member can fail due to flexural stress and fatigue damage and excessive settlement (geotechnical failure). If one of the types of failure occurs, the pavement fails. The failure probability or the reliability of pavement depends on a failure mode. Therefore, it is essential to determine the reliability or margin of safety and corresponding failure probability of pavement taking all the probabilities of pavement failure into considerations.

Probability of failure for fatigue was assessed for RCC pavement. The ASSHTO Guide [39] recommends for reliability analysis the corresponding reliability factor for rigid pavement is between 0.35–0.40.

The analyzed limit state was defined as the failure of the fatigue of the RCC pavement. The limit state function Z is defined as (6):

$$Z = 10^{N_f} - N_{f\_catalog} \tag{6}$$

In this case the limit state function has the form (7):

$$Z = \left(10^{10.25476 - 11.1872(SR)}\right) - N_{f\_catalog} \tag{7}$$

Taking into account the uncertainty of resistance model and effects model, the limit state function has the form (8):

$$Z = \theta_{N_f}\left(10^{10.254/6 - 11.1872(SR)}\right) - \theta_{N_f\_catalog} \cdot N_{f\_catalog} \tag{8}$$

The input basic random variables are shown in the Table 1. The complete results of calculations by FORM and Monte Carlo Simulation methods are shown in Figs. 2, 3 and 4 and in the Table 2.

Table 1. Basic random variables (Source: [8]).

Variable	Distribution	Mean value	Standard deviation
Nf code	Triangular	60000–170000–280000	–
SR	N	0.433	0.01
Θ_{Nf} [-]	LN	1.0	0.05
Θ_{NT} [-]	LN	1.0	0.05

Table 2. The comparison of results calculated by different methods.

Variable	Method/sensitivity	
	FORM simulation method	Monte Carlo simulation
Calculated values of S_o	0.383	0.389
Typical values of S_o according to [39]	0.35–0.40	

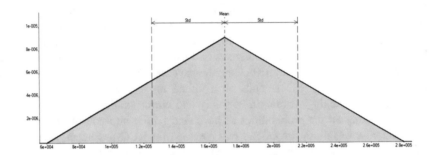

Equivalent axle loads

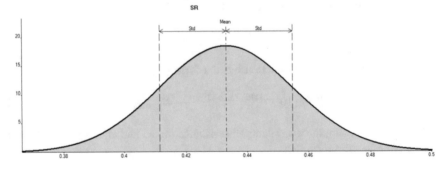

SR

Fig. 2. Histogram and distribution for basic variables

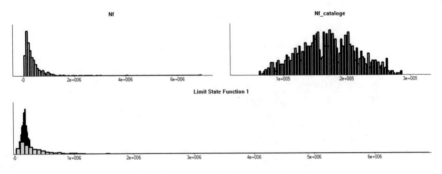

Fig. 3. Histogram for limit state function

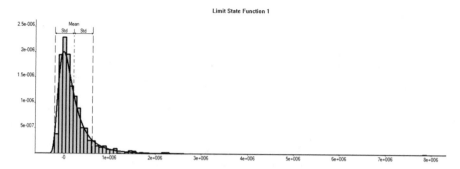

Fig. 4. Histogram and Frechet max distribution for the limit state function

5   Discussion

In recent years, the importance of assessment of structural reliability has increased significantly. This fact is confirmed by the recommendations of the standard PN-EN 1990 [32], JCSS [36], ISO [34, 35] and AASHTO [39]. The AASHTO [39] specifies rules and requirements to ensure safety, serviceability and durability of the pavement. According to the this recommendations, reliability of RCC pavement was calculated. The received values of S_o calculated using semi-probability and probability methods are comparable with the recommended typical values of S_o according to AASHTO, so, the analyzed RCC pavement for the KR 2 traffic category and for the planned axle loads with a triangular distribution with an extreme 170000 is reliable. So, using the probabilistic approach, it was possible to verify a sufficient safety of the RCC pavement.

6   Conclusions

Following conclusions may be drawn from the present studies:

- knowing the distributions and their COV values, the reliability can be estimated for any given set of input,
- using Eq. 8, one can find the R value. That is how a pavement design solution can be obtained,
- significant variability with traffic, necessitates incorporation of probabilistic approach in pavement design,
- a simple pavement design method, where the pavement design solution can be obtained with semi-probabilistic and probabilistic method.

The probability method can be use as a viable tool to deliver reliable pavements than the using conventional deterministic procedure.

References

1. Wu, Z., Mahdi, M., Rupnow, T.D.: Accelerated pavement testing of thin RCC over soil cement pavements. Int. J. Pavement Res. Technol. **9**, 159–168 (2016)
2. Topličić-Ćurčić, G., Grdić, D., Ristić, N., Grdić, Z.: Properties, materials and durability of rolled compacted concrete for pavements. Zastita Materijala **56**, 345–353 (2015)
3. Rupnow, T.D., Icenolge, P.J., Wu, Z.: Laboratory evaluation and field construction of roller compacted concrete for testing under accelerated loading. Proc. Transp. Res. Rec. J. Transp. Res. Board **2504**, 107–116 (2015)
4. Deja, J.: Polish roads. Rev. Road Bridg. Tech. **9**, 3–10 (2003)
5. Czarnecki, L., Woyciechowski, P., Adamczewski, G.: Risk of concrete carbonation with mineral industrial by-products. KSCE J. Civ. Eng. **22**(2), 755–764 (2018)
6. Czarnecki, L., Woyciechowski, P.: Concrete carbonation as a limited process and its relevance to CO2 sequestration. ACI Mater. J. **109**(3), 275–282 (2012)
7. Dobiszewska, M.: Waste materials used in making mortar and concrete. J. Mater. Educ. **39**(5–6), 133–156 (2017)
8. Harrington, D., Abdo, F., Adaska, W., Hazaree, C.: Guide for Roller-Compacted Concrete Pavements. National Concrete Pavement Technology Center. Institute for Transportation, Iowa State University, Iowa (2010)
9. Macioszek, E., Lach, D.: Analysis of traffic conditions at the Brzezinska and Now-ochrzanowska intersection in Myslowice (Silesian Province, Poland). Sci. J. Sil. Univ. Technol. Ser. Transp. **98**, 81–88 (2018)
10. Macioszek, E., Lach, D.: Comparative analysis of the results of general traffic measurements for the Silesian Voivodeship and Poland. Sci. J. Sil. Univ. Technol. Ser. Transp. **100**, 105–113 (2018)
11. Macioszek, E.: Analysis of significance of differences between psychotechnical parameters for drivers at the entries to one-lane and turbo roundabouts in Poland. In: Sierpiński, G. (ed.) Intelligent Transport Systems and Travel Behaviour. AISC, vol. 505, pp. 149–161. Springer, Cham (2017)
12. Mackiewicz, P.: Thermal stress analysis of jointed plane in concrete pavements. App. Therm. Eng. **73**, 1167–1174 (2014)
13. Mackiewicz, P.: Analysis of stresses in concrete pavement under a dowel according to its diameter and load transfer efficiency. Can. J. Civ. Eng. **42**(11), 845–853 (2015)
14. Macioszek, E., Lach, D.: Analysis of the results of general traffic measurements in the west pomeranian voivodeship from 2005 to 2015. Sci. J. Sil. Univ. Technol. Ser. Transp. **97**, 93–104 (2017)
15. Mackiewicz, P.: Finite-element analysis of stress concentration around dowel bars in jointed plain concrete pavement. J. Transp. Eng. **141**(6), 06015001-1–06016001-8 (2015)
16. Sztubecka, M., Bujarkiewicz, A., Sztubecki, J.: Optimization of measurement points choice in preparation of green areas acoustic map. Civ. Environ. Eng. Rep. **23**(4), 137–144 (2016)
17. Szydło, A.: Road pavements made of cement concrete. Theory of dimensioning, implementation. Polski Cement, Krakow (2004)
18. Szydło, A., Mackiewicz, P., Wardęga, R., Krawczyk, B.: Catalog of Typical Structures of Rigid Pavements. Report GDDKiA. GDDKiA, Warsaw (2014)
19. Portland Cement Association: Frost Durability of Roller-Compacted Pavements. Portland Cement Association, Quebec (2004)
20. Bland, S.: Roller-Compacted Concrete Pavements Design and Construction. Pavement Cement Association, Virginia (2000)

21. American Concrete Institute: Guide Specification for Construction of Roller-Compacted Concrete Pavements. American Concrete Institute, Farmington Hills (2004)
22. Beycioglu, A., Gultekin, A., Aruntas, H., Gencel, O.: Mechanical properties of blended cements at elevated temperatures predicted using a fuzzy logic model. Comput. Concr. **20**(2), 247–255 (2017)
23. Skrzypczak,, I., Radwański, W., Pytlowany, T.: Durability vs technical - the usage properties of road pavements. https://www.e3s-conferences.org/articles/e3sconf/abs/2018/20/e3sconf_infraeko2018_00082/e3sconf_infraeko2018_00082.html
24. Skrzypczak, I., Radwański, W., Pytlowany, T.: Choice of road building technology - statistic analyses with the use of the Hellwig method. https://www.e3s-conferences.org/articles/e3sconf/abs/2018/20/e3sconf_infraeko2018_00081/e3sconf_infraeko2018_00081.html
25. Asfal, J.: Difference Between Flexible And Rigid Pavement. http://www.engineeringintro.com/transportation/road-pavement/comparison-between-flexible-and-rigid-pavement-properties/
26. Arangio, S.: Reliability based approach for structural design and assessment: performance criteria and indicators in current European codes and guidelines. Int. J. Lifecycle Perform. Eng. **1**(1), 64–91 (2001)
27. Casas, J.R.: Bridge management: actual and future trends in Bridge Management. In: Life Cycle Performance and Cost. Taylor and Francis, London (2006)
28. Frangopol, D., Das, P.: Management of bridge stocks based on future reliability and maintenance cost. In: Das, P.C., Frangopol, D.M., Nowak, A.S. (eds.) Current and Future Trends in Bridge Design Construction and Maintenance. Safety, economy, sustainability and aesthetics, pp. 45–58. Thomas Thelford, London (1999)
29. Glowienka, S., Fischer, A., Krauss, M.: Implementation of probabilistic method in structural design. In: Proceedings of 9th IPW, pp. 274–286 (2011)
30. Nowak, A.S., Collins, K.R.: Reliability of Structures. McGraw-Hill Higher Education, New York (2000)
31. Sakka, Z., Assakkaf, I., Al-Yaqoub, T., Parol, J.: Structural reliability of existing structures: a case study. Int. J. Civ. Environ. Struct. Constr. Arch. Eng. **8**(11), 1173–1179 (2014)
32. The European Union: EN 1990:2002 Eurocode - Basis of structural design. European Committee for Standardization, Brussels (2002)
33. The European Union: EN 1991-1-7: 2006. Eurocode 1: Actions on structures - Part 1-7: General actions Accidental actions. European Committee for Standardization, Brussels (2006)
34. International Organization for Standardization: ISO 13824:2009 - Bases for design of structures - Part 1-7 General actions. Accidental actions. International Organization for Standardization, Geneva (2009)
35. International Organization for Standardization: ISO 2394: 2015 General principles on reliability for structures. International Organization for Standardization, Geneva (2015)
36. JCSS Probabilistic Model Code Part 3: Resistance Models. https://www.jcss.byg.dtu.dk/-/.../jcss/.../probabilistic_model_code/
37. Prozzi, J.A.: Reliability of Pavement Structures using Empirical-Mechanistic Models. http://www.ce.utexas.edu/prof/Manuel/Papers/ProzziGossainManuel_05-1794.pdf
38. Thongram, S., Rajbongshi, P.: Probability and reliability aspects in pavement engineering. IJLTEMAS **5**(3), 15–18 (2016)
39. American Association of State Highway and Transportation Officials: Handbook for Pavement Design, Construction and Management. American Association of State Highway and Transportation Officials, Washington (2015)

Time Zone Operations in Polish Airports as an Element of Transport System

Dariusz Tłoczyński[✉]

Faculty of Economics, University of Gdańsk, Gdańsk, Poland
dariusz.tloczynski@ug.edu.pl

Abstract. The development of air transport system has increased the importance of transport demands. By purchasing the air transport service, passengers choose carriers who shall meet specific, tailor-made expectations. For passengers, convenient hours of flight operations constitute one of the factors which determine the choice of transport service, whereas for the carrier, it is an important element of air transport management system. Such assumptions provide grounds for conducting analysis on the impact of how time zones of flight operations on the choice of particular carrier. This aim became grounds for research regarding factors affecting the decision making process related to choosing a particular carrier. The results of research conducted by personal interviews with passengers representing various travel segments, regarding passenger preferences constitute grounds for defining the transport service offer. However, in air transport there are some limitations - slots which must be respected by all carriers since they introduce a certain order in the world air transport system. The conducted research and analyses have proved that the times of flight operations and destinations affect the decision on air travel.

Keywords: Air transport · Timetable · Schedule planning · Time zone

1 Introduction

In the 21st century, the world transport service is no longer a luxury but the common good. People make use of transport services rendered by air transport on a mass scale. In 2017, carriers operating on the market carried over 4 billion passengers, which generated the increase by ca. 6%. The data from the past decade and prospects for the future indicate that the upward trend in passenger transport should be sustained.

The development of demand for air transport services results in adjusting the air transport service supply at all levels, i.e. in terms of transport, airport capacity, ground handling, air traffic management (ATM) and in terms of security, implementation of new technologies, impact on air transport passenger preferences and purchasing behaviour.

In order to ensure suitable coordination, in the world air transport system certain regulations are applied assigned to particular regions in the world. It should, first of all, ensure full integration of the external and internal air transport systems. Time zones are the most common instrument applied. By offering transport services, air transport carriers are required to respect the slots. Whereas passengers, expecting high standard of service, choose these carriers who offer convenient departure and arrival times.

© Springer Nature Switzerland AG 2020
E. Macioszek and G. Sierpiński (Eds.): Modern Traffic Engineering in the System Approach
to the Development of Traffic Networks, AISC 1083, pp. 46–55, 2020.
https://doi.org/10.1007/978-3-030-34069-8_4

2 Theory and Methodology

The air transport system is defined in literature related to the transport economics. Janić [1] defines as physical (demand and supply) and non-physical (operating rules and procedures) components. The analysis of transport system constituted the subject of numerous studies by Cascetta [2], Hirst [3], Piskozub [4] and de Neufvillel [5] these authors emphasize the relations between the demand and the supply side of the market at specific resources and production factors within particular economic system. It means that by performing the analysis of transport system one shall indicate two subsystems: macroeconomic subsystem related to economy, resources and legislative policy, and microeconomic - defining the operations of entities on the transport market, including in particular operating resources (infrastructure and suprastructure). Koźlak [6] is right claiming that with reference to transport the character of operating resources in transport is completely different than the one related to operating resources in other sectors of economy.

With reference to air transport the system comprises two basic components: physical and non-physical ones. Physical components include demand and supply. The external and internal consumers of transport system represent the demand component. The system fixed and mobile infrastructure and facilities that produce the services represent the supply component. The non-physical component is represented by a set of operating rules and procedures regulating the operations of physical components to provide safe and efficient transport [7].

The demand for air transport services is reported by the following groups of entities:

- individual passengers,
- enterprises, organisations and institutions sending employees on business trips,
- intermediaries organising group travels - travel and tourist agencies,
- enterprises reporting demand for cargo and mail transport [8].

Moreover, the demand for air transport services may be reported by:

- carriers to aircraft producers,
- airports to carriers and aviation equipment contractors,
- agencies managing air traffic to navigation equipment suppliers.

Certainly, the volume of demand comprises numerous elements. The literature on air transport market indicates passengers' travel preferences and behaviour (i.e. type of travel and destination, frequency, financial status and income level of passengers, etc.), number of passengers making use of air transport system, prices of competitors and related travel expenses and passengers' expectations with respect to future prices [9]. As for carriers, it refers to the type and size of aircraft, seat capacity, etc. Moreover, other factors affecting the development of demand include: level of technique and technology, own assets, availability of capital, human and tangible resources, cultural level, habits, traditions, customs as well as fashion or social pressure and expectations.

Whereas, the supply of transport services is represented by carriers at various configurations. The entities operating on the air transport service market on the supply side also include airports, handling agencies, aviation equipment contractors, aircraft producers, ATM, etc. Their offer depends on numerous factors. The most important of them include:

- accessibility to infrastructure and mobility of devices,
- service selling channels,
- airport ground access systems,
- airport infrastructure,
- airspace density,
- facilities and equipment for processing information,
- legislation (internal and external) and regulations.

Therefore, within the transport system the demand meets the supply related to air transport services at various possible factors defining the air transport environment. In the case of air transport services the subject of trade is not the service itself but first of all, factors defining the transport service.

3   Timetable as an Element of Transport Service Offer in Polish Airports

In Poland, there are 15 airports (1 central airport - Warsaw Chopin Airport, and 14 regional airports). The Polish airports offer regular service to several hundred destinations in the world (Table 1) and charter service to attractive tourist resorts where the flight is ordered by tourist agency or tour operator, as well as General Aviation services. Moreover, apart from passenger transport, the Polish airports provide cargo and mail transport service.

Table 1. Number of scheduled services and number of carriers operating in the Polish airports in 2019.

Airport	Number of final destination			Number of carriers		Number of air networks per week
	Domestic	Intra Europe (without Poland)	Intercontinental	Low Cost Carrier	Legacy Carrier	
Bydgoszcz	–	8	–	1	2	28
Gdańsk	3	61	3	4	7	397
Katowice	1	38	1	3	2	211
Kraków	3	85	2	10	10	541
Lublin	1	5	–	3	1	25
Łódź	–	5	–	1	1	16
Modlin	–	47	1	1	–	184
Olsztyn	–	3	–	2	–	7
Poznań	1	29	–	2	3	146
Radom	–	–	–	–	–	–
Rzeszów	1	9	1	1	2	77
Szczecin	1	8	–	3	2	42

(*continued*)

Table 1. (*continued*)

Airport	Number of final destination			Number of carriers		Number of air networks per week
	Domestic	Intra Europe (without Poland)	Intercontinental	Low Cost Carrier	Legacy Carrier	
Warszawa	9	78	19	3	23	1709
Wrocław	2	53	2	3	4	253
Zielona Góra	1	1	–	–	1	6

On the Polish market, there are tens of air carriers offering regular air services. In 2018, low cost carriers had a market share of nearly 60%. Particularly important are the first five of them (TOP air carriers in Poland: Ryanair, LOT Polish Airlines, Wizz Air, Lufthansa and EasyJet) [10]. These carriers had a market share of over 80%, which is indicative of significant concentration in the sector of passenger transport.

Upon choosing air transport carriers, passengers have numerous travel demands. The most frequent expectations include: short door-to-door travel time compared to other sectors of transport, direct connections, convenient departure and arrival times (flight operations), ticket price and available connection network (see Fig. 1).

One of the instruments used to develop the supply of service in air transport is the flight timetable. The literature on air transport economics fails to define the concept of flight timetable. It must be indicated that it is a scheduled and publicly available mode of performing commercial flights, taking into account infrastructure, time necessary for operations and other factors inherent to this situation.

From the perspective of carriers and infrastructure managing entity, the flight timetable constitutes the transport scheme, whereas for passengers it constitutes the transport company offer. The most important elements affecting the timetable include:

- no. of flight,
- air carrier (operator),
- departure time,
- arrival time,
- date of operation,
- departure airport,
- arrival airport,
- type of aircraft.

The timetable planning is affected by numerous factors:

- airport accessibility - hours of flight operations, duration of stay at airports, possibility to base aircrafts, frequency of rotations, adjustment of hours the needs of particular segment. variety of airport locations with numerous arrival and departure limitations, as well as the number of gates, ground personnel and equipment [11],
- airspace capacity - possibility to handle airspace via ACC (Area Control Centre), APP (Approach Control), TWR (Tower) and in some airports, GND (Ground Controller) [12], working hours, number of employees,

- slots and separations - airport slots ensure carriers the right to use airports within particular hours [13]. The slots are used when airports have certain limitations, e.g. as for runway infrastructure capacity [14] (as in the case of e.g. Heathrow Airport) or parking area availability (as in the case of many airports at Greek islands). Slots can be traded. If the volume of air sector is lower than the demand, the sector must be limited. It is related to the security policy. The overload of air traffic controller resulting from too large number of aircrafts assigned to the controller at the same time is simply too hazardous, in particular when in the sector there is complex traffic of aircrafts at intersecting routes where the cruising level is increased or decreased. Whereas, separation refers to the time between consecutive flight operations [15],
- managing the connections network - decision on the choice of markets to operate on, decision on the choice of routes, airports, type of connections network, decision on the number of air services, choice of connection types, availability and type of aircrafts used and frequency of performed operations, various aircraft operational characteristics, their cost-based specificity, transport capacity and seasonal maintenance requirements imposed by certifying authorities, number of destination airports, potential routes and handling those origin-destination (O-D) markets which differ in the demand volume and passenger travel behaviour,
- availability of air crew and ground personnel - management of working time related to air connections network, management of working time related to timetable, management of personnel related to aircraft model, working time limitations, air crew logistics in extraordinary situations, e.g. accommodation, change of crew, auxiliary crew in other airports [16, 17].

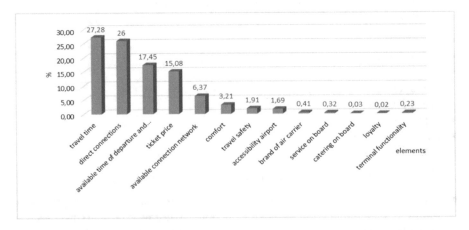

Fig. 1. Expectations of passengers making use of Polish airports in 2013 (Source: own marketing research conducted at airports)

Certainly, the optimization of timetable planning covers the design of aircraft use and crew turnover schedule. Its aim involves defining the schedule for generating profit coherent with the operational, marketing and strategic goals.

4   Time Zones in Airports

An important factor affecting the efficiency of aircraft use involves time zone operations at airports. Air transport operates based on UTC universal time, also named
ZULU - marked with letter Z, e.g. 14:35 Z or 14:35 UTC - for the aviation and tourist
sector both markings are equivalent in meaning. In air transport, flight planning,
operational schemes, documents - flight schedule, information on flights, departure,
flight duration, arrival, in meteorology - current weather and forecast, UTC time is
applicable. However, the time is always referred to local time for a particular airport (as
per correct time zone). In Poland, UTC+ 2 h is applicable. It does not have any effect
on the correct operation of air transport system. Air travels frequently take place
between time zones and in such case UCT time is regularly converted to the local
departure time (crucial for consumers and passengers). At the points of transport,
including navigation points located at flight routes, air crew makes use of UTC time
and does not convert it to local time since it is unnecessary. Whereas, for passengers, it
is more convenient to use local time related to the destination time zone within a
particular area.

   Although in air transport the official time related to flight operations includes hours
as per UTC, in airports local times are common for passenger service.

   Based on the management of the so-called time space, at airports there are time
blocks in the form of slots determined for performing particular flight operations
(Fig. 2).

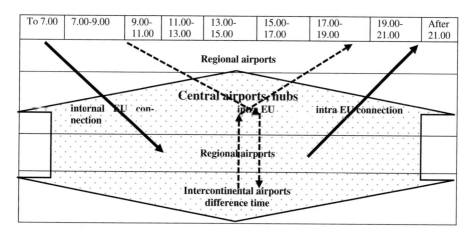

To 7.00	7.00-9.00	9.00-11.00	11.00-13.00	13.00-15.00	15.00-17.00	17.00-19.00	19.00-21.00	After 21.00

Fig. 2. Time slots at Polish airports at services within the EU and at intercontinental routes

   The connections from regional airports to other regional, local airports are performed by carriers within the airport working hours taking into account the above
mentioned assumptions related to safe flight operations at airports, whereas services at
intercontinental routes must be adjusted to particular time zones. The average number
of flight connections on an hourly basis for Warsaw Chopin Airport and regional
airport i.e. Gdańsk Airport is presented in Fig. 3.

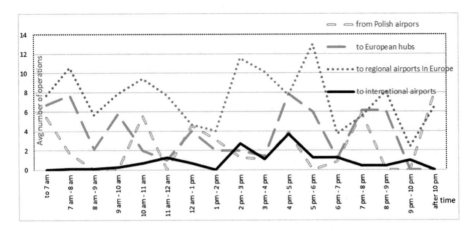

Fig. 3. Average number of selected flight operations at Warsaw Chopin Airport on an hourly basis (Source: [18, access data: 15.04.2019])

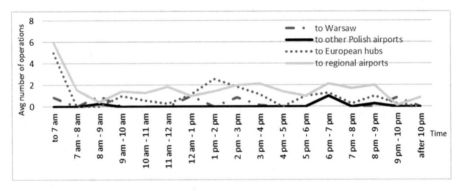

Fig. 4. Average number of selected flight operations at Gdańsk Airport on an hourly basis (Source: [18, access data: 15.04.2019])

The presented data reveals several dependencies. There are connections between regional airports where business travels are handled. The hourly accessibility of flight services provides passengers with a possibility to travel (O-D, O-D) within one day; for entrepreneurs it is undoubtedly a significant benefit related to saved time and financial resources dedicated to business trips, hotels and alternative costs. Employees who comes back earlier from business trip can come to work the following day (Fig. 4).

Such connections are exemplified by trips from regional airports to London; at larger regional airports carriers offer several connections per day at different hours.

Whereas, connections from regional airports via main hub airports are handled by carriers within such time zones to enable passengers quick transfer to other airports located on the continent (min. transfer time at hub airport is scheduled at ca. 50–60 min.) or airports outside the continent (transfer time amounts to min. 90 min.). Example hours of start and finish of travel are presented in table below (Table 2).

Table 2. Indirect connections from Gdańsk Airport to Rome Airport.

Origin (departure time)	Destination (arrival time)	Via/Air carrier	Transfer time	Total time
Gdańsk	Rome			
6:10	12:35	Copenhagen/SAS	2:55	6:25
6:15	12:35	Frankfurt/Lufthansa	2:45	6:20
6:25	11:50	Amsterdam/KLM	0:50	4:50
13:10	20:00	Oslo/Norwegian	2:10	6:50

Passengers travelling via hub airports to the European destinations most frequently start their journey in the morning until 7.00 am (50% of respondents taking continental flights with the use of transfer). Attention should be paid to the network of links to the airport in Warsaw between 10.00–12.00. The main passengers at these links include people traveling with PLL LOT from Warsaw at intercontinental routes.

The passengers surveyed in the Polish airports considered convenient departure and arrival times as significant (17%) upon selecting a particular carrier. The demand for convenient time zones was considered most important by 14% of the respondents in the Polish airports. Whereas, among passengers traveling at:

- national routes - by 9%,
- direct European routes - by 13%,
- European routes with the use of transfer in hub airports - by 20%,
- intercontinental routes via Warsaw - by 12%,
- intercontinental routes via European hub airports - by 25%.

Whereas, taking into account the proportion and structure of traffic in airports we can determine the segments where departure and arrival times are most important (Fig. 5).

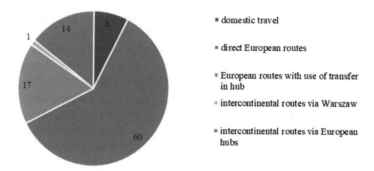

Fig. 5. Structure of flights where convenient departure and arrival times are most important

At the Polish airports the surveyed passengers for whom the most important factor upon selecting the carrier involved convenient departure and arrival times most frequently travelled at direct routes (60%), more frequently from regional airports (33%) than from Warsaw Chopin Airport (27%). Convenient time zones were least important, as the most important factor upon selecting the carrier, for passengers travelling at continental routes from regional airports with the use of transfer in Warsaw (Fig. 6).

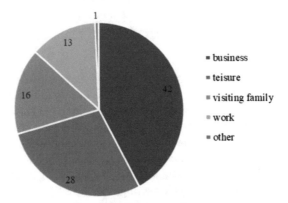

Fig. 6. Structure of passengers departing from the Polish airports for whom the most important factor upon selecting the carrier were convenient departure and arrival times

The largest group of passengers for whom the most important factor upon selecting the carrier were convenient departure and arrival times included passengers traveling on business (42%), for leisure (28%), whereas passengers traveling to work and to meet the family constituted less than 16%.

5   Conclusions

The hours of performing flight operations at airports result from several aspects. Certainly the most important element involves the role of airport in the transport system. Despite the fact that airports can be divided as per various classifications, each airport performs the same functions within the transport system. Airports perform social, promotional and economic functions related to the development of the region as well as integration-related functions. The latter, in particular, are performed by airports in cooperation with carriers by adjusting the hours of flight operations.

From the operational perspective, the relations between airport managing operators and carriers focus on providing passengers in a particular region with comprehensive and broadest air transport offer.

In the world, particular groups of routes are combined to separate, for particular directions, special time zones for operations performed in airports. For example, in the morning passengers are carried to the European hubs, so that they can take the offer of carriers and travel to the European airports. But at the same time, at some routes passengers can return to their primary place of departure. Such solutions are extremely important and desirable by passengers taking transfer trips. Most frequently, these passengers travel on business.

The presented research results indicate that studies on passengers' travel behaviour with carriers' capabilities, performance of airports and entities managing the airport infrastructure should constitute the subject of scientific research and application-type research for air transport entities.

References

1. Janić, M.: Air Transport System Analysis and Modelling. Transportation Studies 6. Gordon and Breach Science Publishers, London (2000)
2. Cascetta, E.: Transportation System Analysis. Models and Applications. Springer, New York (2009)
3. Hirst, M.: The Air Transport System. Woodhead Publishing Limited, Cambridge (2008)
4. Piskozub, A.: Gospodarowanie w transporcie. Podstawy teoretyczne. WKiŁ, Warsaw (1982)
5. De Neufville, R., Odoni, A.: Airport Systems. Planning, Design and Management. McGraw-Hill, New York (2003)
6. Koźlak, A.: Ekonomika Transportu. Gdańsk University Press, Gdańsk (2007)
7. Janić, M.: Air Transport System Analysis and Modelling. Capacity, Quality of Services and Economics. Gordon and Breach Science Publishers, London (2000)
8. Rucińska, D., Ruciński, A., Tłoczyński, D.: Transport lotniczy. Ekonomika i organizacja. Gdańsk University Press, Gdańsk (2012)
9. Wensveen, J.G.: Air Transportation. A Management Perspective. Ashgate, Farnham-Burlington (2007)
10. Civil Aviation Authority (CAA) in Poland. https://www.ulc.gov.pl/en/270-english/current-information/3838-entities-supervised-by-civil-aviation-authority-of-the-republic-of-poland-infographics
11. de Neufville, R., Odoni, A.: Airport Systems. Planning, Design and Management. Aviation Week, Washington (2003)
12. Procedures for Air Navigation Services - Air Transport Management (PANS-ATM), Doc. 4444, 16th edn. (2016). https://ops.group/blog/2016-16th-edition-icao-doc-4444/
13. IATA, Worldwide Scheduling Guidelines, 13th edn. (2006). http://www.wwacg.org/up/files/docsWSG/WORLWIDE_SCHEDULING_GUIDELINES/WSG_13th_Edition.pdf_040309_032641.pdf
14. Gilbo, E.P.: Airport capacity: representation, estimation, optimization. IEEE Trans. Control Syst. Technol. 1(3), 144–153 (1993)
15. Cheng-Lung, W.: Airline operations and Delay Management. Ashgate, Farnham-Burlington (2010)
16. Belobaba, P., Odoni, A., Barnhart, C.: The Global Airline Industry. Wiley, Hoboken (2016)
17. Wells, A.T.: Air Transportation. A Management Perspective. Wadsworth Publishing Company, Belmont (1999)
18. Flight Radar. http://www.flightradar24.com

Actions for Reduction of the Environmental Impact of Public Transport in the Górnośląsko-Zagłębiowska Metropolis

Bartosz Bociąga[✉]

Faculty of Transport, Silesian University of Technology, Katowice, Poland
bartosz.bociaga@polsl.pl

Abstract. The paper discusses the strategy of Górnośląsko-Zagłębiowska Metropolis (GZM) for the reduction of environmental impact of public transport operators. The evaluated environmental impact of the public transport is substantial and constitutes a serious health hazard for the citizens of the Metropolis. GZM carries out diverse actions for the reduction of these threats and promotes initiatives which involve not only public transport operators but also users of their services. The actions are performed in compliance with national and European Union (EU) environmental directives and regulations. Important for the success of these actions are finance sources especially EU funds. The most important action is the promotion of electro mobility that is elimination of diesel bus fleet and funding the purchase of electric buses and extending the charging infrastructure.

Keywords: Pollution · Public transport · Electromobility · Environmental programmes

1 Introduction

Increasing transport intensity leads to increased risks associated with environmental pollution. This is closely related to climate change and the loss of biodiversity. Currently undertaken activities are aimed at counteracting these tendencies, or at least reducing the pace of their growth. Technological improvements contribute to reducing the degree of air pollution caused by road transport, despite the increase in traffic volume. Nevertheless, solving the problem of environmental pollution in urban areas requires additional actions.

The main pollutants are substances emitted by vehicle engines due to combustion of fuel. In the combustion process, the chemical energy of the fuel is converted into mechanical energy. The combustion process involves the production of large amounts of harmful gases, including nitrogen oxides (NOx) and solid particles. Pollutants emitted during the combustion of liquid fuels in automotive vehicles affect the processes of acidification of the environment and the ground ozone production. Despite the use of various types of mechanisms to reduce the harmful effects, it is almost impossible to completely purify the exhaust gases, and thus neutralize their negative impact on the natural environment. Important constituent of pollution are also particles of gum and asphalt raised into the air by vehicle tyres in the course of driving.

© Springer Nature Switzerland AG 2020
E. Macioszek and G. Sierpiński (Eds.): Modern Traffic Engineering in the System Approach to the Development of Traffic Networks, AISC 1083, pp. 56–65, 2020.
https://doi.org/10.1007/978-3-030-34069-8_5

European Union (EU) is a legal body issuing directives on the limitation of air pollution - EURO. Euro is a European emission standard of permissible exhaust emissions in new vehicles sold in the European Union. These standards were developed in a series of European Directives that successively increased their stringency. Every few years new, increasingly stricter emission standards are introduced. Euro 1 was introduced in 1992 - that is over 25 years ago. Currently Euro 6 is in effect and vehicle producers have no problems with adapting their products to the applicable requirements [1].

There are also many European Union directives and regulations defining the principles of monitoring and reporting greenhouse gas emissions, harmful gases and exhaust emissions. Poland is a signatory of international agreements and conventions, among others [2]:

- United Nations Convention on Climate Change and the Protocol from Kyoto,
- Convention on Biological Diversity and Protocol on Biological Safety,
- Vienna Convention for the Protection of the Ozone Layer and the Montreal Protocol on a Substance that Depletes the Ozone Layer,
- Basel Convention on the Control of Trans Boundary Movements of Hazardous Wastes and Their Disposal,
- Convention on International Trade in Wild Plants and Animals species threatened with extinction,
- Convention on the Protection of the Marine Environment of the Baltic Sea Area.

The introduction of increasingly stringent standards and the need to improve fuel efficiency have significantly contributed to the technological development of the European automotive industry and transport. Innovations in this area include ecological solutions, improvements in the technology of conventional engine and exhaust system as well as development of electric and hybrid vehicle technologies. In the public transport, over the last few years, a dynamically progressing process of exchange of urban bus fleets has been observed. It is based on, low-emission technologies in public urban transport.

Poland as a member of the European Union must take into account its legal provisions. The current environmental regulations of the European Union include about 200 documents that relate to water and air pollution, waste management and chemical substances, biotechnology. The Polish Committee for Standardization, is adopting existing European standards in the field of environmental protection (there are about 9,500 of them), cooperates with other European Union countries to use their experience in the implementation of environmental regulations. The Polish system of financing the ecological investment is based on the Parliament Act of April 27, 2001 - Environmental Protection Law. Its essence is the integration of many sources of investment financing for environmental protection. It consists of [2]:

- state budget and budgets of local government units,
- ecological funds, business entities,
- banks, foundations and agencies,
- foreign institutions.

The paper is composed of 4 section. In the first one, the introduction is about the current situation of air conditions in the agglomeration GZM and legal body issuing directives to reduce the pollutants. In the second section, environmental impact of public transport and health hazard caused by local operators have been presented as a summarized value of operator's fleet, average length of transport journeys and emitted pollutions. In the third section, have been discussed the eventually possible actions to reduce the pollution by public transport operators, such as: increasing the attractiveness of public transport, integrated transport, introduction of gas or electric transport vehicles. And fourth section is conclusion.

2   Environmental Impact of Public Transport and Health Hazards Caused by Pollutants

The current environmental impact of public transport used in GZM by regional operators is summarized in Table 1. It is based on the information from public transport operator's fleet, the average length of transport journeys in 2017 year and Euro emissions standards. Lack of the split of travel lengths into vehicle categories requires an assumption on the way the companies fleets are exploited in order to assess the compound pollution effect. Available reports indicate that companies do not assign vehicles to bus lines according to their Euro emission levels. Vehicles in working order are dispatched to service bus lines daily, which accounts usually to more than 80% of the company's fleet. Taking this into account, on average, each vehicle in service, travels yearly about eighty thousand km.

Table 1. Emissions of pollutants by the fleet of transport companies in the Górnośląsko-Zagłębiowska Metropolis.

Emitted substance [kg/km]	Euro 1	Euro 2	Euro 3	Euro 4	Euro 5	Euro 6
CO	0	2 138	9 330	2 938	13 181	36 468
HC	0	321	875	294	1 318	3 695
NOx	0	1 176	7 289	1 469	4 745	16 100
HC+NOx	0	1 497	8 164	1 763	6 063	20 506
PM	0	171	729	53	132	1 173
Total	0	5 302	26 388	6 517	25 439	77 943

The sums are split into categories of vehicles used to carry out transport work. The largest share of pollutants is emitted by Euro 3 vehicles. This standard was introduced 18 years ago. It indicates that the average age of these vehicles servicing public transport is at least 18 years. Presumably these vehicles will go out of service in near future. Euro 4 vehicles contribute to a similar extent as very old Euro 2 vehicles. New vehicles complying to Euro 5 approach the levels of pollution of the Euro 3 vehicles but the fleet is almost 63% larger. The contribution of Euro 4 vehicles is four times smaller than that of Euro 3, this can be accounted to a very small number of these vehicles, which may reflect some change in the vehicle replacement policies of the companies a few years ago.

The sum of pollutants of Euro 5 and 6 vehicles gives 51% of the total pollution although these add up to 65% of the total number of vehicles. This shows that companies care for the reduction of pollution. There are incentives for such a behaviour. For instance, companies can apply for subsidies from environment protection bodies such as National Fund for Environmental Protection and Water Management. Tender documents, in the case of buying vehicles for public transport, must include terms of reference which contain clauses that restrict the pollution levels of the vehicles.

The exposure to pollutants causes in the largest extent dysfunctions of respiratory, cardiovascular and immunologic systems. It is the source of toxicity which leads to a variety of cancers in the long term. Many factors affect the health condition, including the level and availability of health care, lifestyle or the level of affluence of the society and additional factors on which there is no direct influence such as: modifiable environmental factors, such as air quality, noise, radiation, green areas, i.e. those whose quality or level impacts depends largely on human activity. Transport contributes to the degradation of the natural environment and has a negative impact on the people. In the European Union, it is a source of almost 54% of the total emissions of nitrogen oxides, 45% carbon monoxide, 23% non-methane volatile organic compounds (NMVOC) and 23% PM10 dusts and 28% PM2.5 dusts, it is also responsible for over 41% of ozone precursor emissions troposphere and 23% of CO_2 emissions and almost 20% of other greenhouse gases [3]. In recent years, 30% of decrease is noted for CO_2 emissions by newly registered cars [4]. CO_2 emissions decrease when the quality level of road infrastructure increases, a good consistency road segment has noticed an emission rate 20–30% lower than a poor-consistent one [5].

The first effect of pollutants is voice disturbances due to damages of human respiratory tracks. Pollutants favour the development of asthma and lung cancer, especially PMs and hydrocarbons cause serious damage to the respiratory tract. Studies have shown that traffic-related air pollution increases the risk of chronic obstructive pulmonary disease (COPD) [6].

Exposure to pollutants can lead to changes in white blood cell counts, which affects the cardiovascular functions. High levels of NO_X, are associated with ventricular hypertrophy. Air suspended toxic materials have damaging effect on the nerve system. Reports reveal that pollutants increase the incidence rate of neuroinflammation, Alzheimer's and Parkinson's diseases. There is also evidence that neurobehavioral hyperactivity is increased.

Immune system dysfunction brings about an increased risk of numerous diseases. Air pollutants modify antigen presentation, increase the serum levels of the immunoglobulin which decreases the resistance to illnesses.

Cancer risk caused by air pollutants is severe. Lung cancer is the dominant disease each year, about 19,000 new cases are reported in Poland. Based on the analysis prepared by Health and Environmental Alliance, it follows that every 8th lung cancer case in Poland is caused by air pollutants [6].

In 2016, malignant tumours were the second cause of death in Poland, causing 27.3% of deaths among men and 24.1% of women. The incidence of malignant tumours in the Silesia region in 2016 was: 9852 for men for every 100,000 inhabitants and 9704 for women for every 100,000 inhabitants [7].

The amount of average annual PM10 dust concentrations has slightly decreased in recent years. It is a visible decrease in concentrations at sites in Gliwice or Sosnowiec. Many factors influenced this condition, including meteorological conditions, as well as the activity of both organizational units and numerous remedial actions taken by local governments or the new fleet of local transport organisations [17].

3 Actions for Reduction of Pollution by Public Transport Operators

Undertaken actions related to the reduction of the negative impact on the environment are of a national, transnational or even global nature. As part of global processes with negative environmental aspects, climate change, environmental pollution, and environmental imbalance are associated. The nationwide processes are closely related to local processes, such as: increasing awareness among agglomeration residents, strengthening market institutions in relation to environmental values, the ability to find new functions for environmentally valuable areas, tourism development, increase in prices of natural resources and increase in value of protected areas (Parks National, environmental objects) [2].

In Poland, the responsibility for the natural environment of both state and self-government authorities and business has increased, which is reflected in the reduction of pollution and the development of ecological funds. The implementation of the principles of sustainable development forces the creation of many institutions that will set goals and organise the means to implement them.

The major projects in the GZM aimed at reducing the negative impact on the environment of transport activities are [8, 9]:

- increasing the attractiveness of public transport,
- integration of transport means (bus, tram),
- introduction of gas or electric transport vehicles.

All of these projects allocate sources to the development of:

- efficient methods for coordinating the composition of budgets for financing tasks of the projects,
- mechanisms for the analysis of technical cooperation related to the transfer of technology and know-how between entities taking part in the projects,
- methods for assessments of the impacts of international regulations in the field of sustainable development,
- methods for effective participation in the process of creating international regulations in the field of promoting environmentally friendly actions,
- procedures for resolving international disputes in the field of sustainable development.

Programs for the development of public transport based on a modern low-emission bus fleet and creation of an integrated public transport system (tram/bus/train) are proposed. Many programs are aimed at changing the attitudes of private car users towards acceptance of public transport. The guidelines for the programs are presented in the documents: "The strategy of activities for KZK GOP for the years 2008–2020", in the directives published by Śląski Urząd Wojewódzki in Katowice and in the directives published by GZM [10]. In recent years, significant sales of electric vehicles have been recorded. This is due to various actions organized by the European Union. For instance, European Union supports the use of renewable electricity and smart charging; helping to develop and standardise charging infrastructure; and supporting research on batteries. On the local and regional level - the introduction of lower taxes or the provision of free public parking for electric vehicles) are also promoting electric mobility.

3.1   Increasing the Attractiveness of Public Transport

The current strategy adopted by GZM defines programs for the development of public transport based on a modern low-emission bus fleet and creation of an integrated public transport system (tram/bus/train). The aim is to promote the use of alternative means of transport, to private cars, and reduce the entry of private vehicles to congested city centres. GZM is also involved in programs for the enlargement of fleets of electric vehicles. This is coincident with the directives of EU Transport Commissions [11].

Modern technologies provide a strong impetus to improve the management processes of large transport systems and also contribute to the positive perception of developed services.

Local operators in GZM offer a number of services promoting public transport. MZK Tychy offers the system which enables purchasing of tickets using a mobile application – moBilet. The application also offers the bus timetable updated in real time. Another application "Kiedy przyjedzie?", presents the location of buses and informs the passengers when the bus will arrive. It also generates various types of statistics and data for the local operators [12].

Jaworzno is one of the first towns in Poland which introduced a magnetic card instead of "paper" tickets. It is a multifunctional electronic travel card. The card has the form and size of a standard plastic ATM card and it is a contactless card. All buses belonging to the PKM Jaworzno company are equipped with the Open Payment System Jaworzno ticket terminal, enabling the use of the electronic travel cards. Mobile applications for public transport in Jaworzno use the website - kartamiejska.pl which enables the purchase of travel cards using online payment systems [13].

Municipal administrations in GZM launched in several towns Park and Ride systems to encourage users to park their car outside town centres and continue travel using public transport.

3.2   Integrated Transport

Integration of transport means - integrated transport more efficient utilisation of transport resources which in consequence diminishes the environmental impact. In Górnośląsko-Zagłębiowska Metropolis integration of public transport services is the first on the list of strategic goals. Currently the core of services is coordinated by

KZK GOP, MZK Tychy, MZKP Tarnowskie Góry, PKM Jaworzno. Smaller operators are encouraged to join, this expands common transport connections in the Metropolis extending the range of travel. Better connections reduce travel times and contribute to greater attractiveness of public transport, greater ease in travel planning.

Integration means also the introduction of information systems and technologies for streamlining the functioning of transport as part of the Metropolis economy –integrated public transport is being prepared for the use of the Silesian Public Services Card (Śląska Karta Usług Publicznych (ŚKUP)) for making payments for transport services [10].

3.3   Introduction of Gas or Electric Transport Vehicles

The action that gives the largest effect in the domain of reduction of pollution is the transition from the use of diesel based vehicles to vehicles powered by gas or electricity. Such vehicles are named "green". The alternative power sources are at an early stage of development and require higher investments to meet users requirements. In the case of buses there are four powering solutions that is: gas, hybrid, fuel cell and battery.

Hybrid buses do not require any special infrastructure and can operate in the current infrastructure technology. Fuel cells predominately use hydrogen for charging, which is a very hazardous gas and only a few companies have the technology for its safe use. Electric buses positively differ from conventional means of transport [14]:

- the operating costs of electric buses are lower than for conventional buses,
- less noise emission,
- high efficiency of electric drive and lower failure rate,
- no emission of harmful substances.

Electric buses also have disadvantages:

- short range of operation due to limited battery capacity,
- significantly higher costs of purchase,
- require efficient charging infrastructure and batteries services,
- higher susceptibility to changing weather conditions.

The use of electric buses is influenced by:

- terrain - uphill drive requires significantly more energy which results in shortening of the buses range of operation,
- climatic conditions - winter, extra energy is used to heat the vehicle, in summer energy is used for cooling,
- traffic - higher energy consumption due to the low speed, frequent stopping, engine starting.

The basic limitation of the use of electric buses in public transport is their range, and therefore an appropriate charging infrastructure is necessary for sustaining their operation. It is necessary to plan breaks in their operation for charging bus batteries. Two basic solutions are used: slow charging which is done at the operators' depots, or fast charging done at special charging stations located at crucial points of the bus routes network [15].

The introduction of "green vehicles" is warmly welcomed by users who are environmentally conscious. Studies show that Jaworzno residents rate the action of introducing electric buses by PKM Jaworzno at 4.9 on a five-point scale. 93% of Jaworzno residents are happy that the city is investing in electric buses. The first electric fleet Solaris Urbino in the PKM Jaworzno fleet has already travelled over 100,000 kilometres, although it is less than 1.5 years old [16].

Transport in Poland is responsible for around 13% of gas emissions, although there are cities in which the rates reach even 60%. The introduction of environmentally friendly vehicles requires large investments. Funds for activities in this area can be obtained from local and national sources as well as EU subsidies.

Bus purchase thanks to EU funding are notable in each of the Silesian regional operators. Since October 2017, PKM Jaworzno started the operation of 22 electric buses from the Solaris Company. New vehicles in the fleet are: Urbino 8.9 LE electric (4 buses), Solaris Urbino 12 electric (9 buses) and Solaris Urbino 18 electric (9 buses). This means that the operator has 23 electric buses, including the Solaris Urbino 12 electric purchased in 2015. The buses are charged using pantograph chargers or plug-in stationary chargers. The fleet is to be extended in 2020 by further 20 buses and this will bring the share of electric buses to 80% of the whole fleet. The company will also develop its charging facilities with 5 new pantograph charging stations and 10 plug-in chargers that will be mounted in the companies depot [10, 13, 14].

Other GZM operators also commission electric buses using EU funds. PKM Katowice plans to buy 20 electric buses. The tender for the purchase of the first batch - 10 electric buses - was announced in 2017.The range of these buses will be from 50 to 70 kilometres. Charging facilities will be built in 3 Stawy Shopping Center and in city centre near the PKM Katowice office. PKM is planning next purchases in 2019 and 2020.

4 Conclusions

In the Górnośląsko-Zagłębiowska Metropolis carbon monoxide has the largest share in the volume of emissions. Other harmful to human health and the environment, emitted substances are: nitrogen oxides, hydrocarbons, solid particles (PM). The level of pollution generated by public transport is high and dangerous for the health of GZM residents, but companies are aware of the risks and take actions in this area.

The European Union's policy and local policy consistently strive to develop sustainable transport, including sustainable urban mobility. An important direction of changes is the systematic reduction of the negative impact of the transport system on the environment. Direct reflection of these activities is the adoption by GZM of the metropolis of the Directive of the European Parliament and of the Council 2014/94/EU on the development of alternative fuels infrastructure.

In response to EU regulations, the Ministry of Energy has developed the Electro mobility Development Program, which includes [10]:

- Electro mobility Development Plan "Energy for the future",
- Document establishing the Low-Carbon Transport Fund,
- Document establishing electro mobility and alternative fuels.

The most important for the GZM is the National Framework for Alternative Fuels Policy. Its main goal is - by 2020 year 50,000 electric vehicles, 6000 publicly available points with normal charging capacity and 400 high-power points, 3000 CNG vehicles and 70 CNG charging points, and the scope of its activities include the cities of the Silesian agglomeration.

Actions to reduce the emitted environmental pollutions based on: the promotion of electro mobility, the obligation to replace vehicles used by public administration for electric vehicles, the obligation to build adequate charging infrastructure near public buildings, introduction of tax facilities for users of electric vehicles.

Thanks to the introduction of increasingly stringent standards, emissions of bus fleets are gradually reduced. This is confirmed by the high share of new low-emission buses and other activities, including the purchase of electric buses, the creation of joint strategies and plans for emission reduction.

Year after year, the GZM regions transport sector has achieved significant reductions in emissions of some of the major air pollutants - primarily due to the introduction of emission standards, financial measures and, to a lesser extent, due to alternative fuels and avoiding transport activities.

References

1. DieselNet, https://www.dieselnet.com/standards/eu/ld.php
2. Zysnarska, E.: Znaczenie funduszy ekologicznych dla bezpieczeństwa inwestycji w ochronę środowiska. Ekonomia i Prawo **2**, 339–355 (2006)
3. Badyda, A.: Environmental impact of transport. Science **4**, 115–125 (2010)
4. Helmers, E., Leitao, J., Tietge, U., Butler, T.: CO2-equivalent emissions from European passenger vehicles in the years 1995–2015 based on real-world use: assessing the climate benefit of the European diesel boom. Atmos. Environ. **198**, 122–132 (2019)
5. Llopis-Castello, D., Camacho-Torregrosa, F., Garcia, A.: Analysis of the influence of geometric design consistency on vehicle CO2 emissions. Transp. Res. Part D: Transport Environ. **69**, 40–50 (2019)
6. Health and Environment Alliance. https://www.env-health.org/IMG/pdf/13.12.2017_-_boosting_health_by_improving_air_quality_in_the_balkans_ied_briefing.pdf
7. Wojciechowska, U., Czaderny, K., Ciuba, A., Olasek, P., Didkowska, J.: Cancer in Poland in 2016. Studio Mediana, Warsaw (2016)
8. Online System of Legal Acts. http://prawo.sejm.gov.pl/isap.nsf/DocDetails.xsp?id=WDU20170000730
9. Public Information Bulletin of the Górnośląsko-Zagłębiowska Metropolis. http://bip.metropoliagzm.pl/archiwum/article/uchwala-nr-xii-ukosnik-71-ukosnik-2018-zgromadzenia-gornoslasko-zaglebiowskiej-metropolii.html

10. The strategy of activities for KZK GOP for the years 2008–2020. https://www.kzkgop.com. pl/public_media/fb/files/strony/strategia/strategia_kzk_en.pdf
11. Official Journal of the European Union. https://publications.europa.eu/en/publication-detail/ -/publication/d414289b-5e6b-11e4-9cbe-01aa75ed71a1/language-en
12. MoBilet General Information. http://www.mobilet.pl/
13. Karta Miejska General Information. https://kartamiejska.pl/pl-pl/jaworzno/kupbiletjkm. aspx?city=537
14. Krawiec, K.: Proces wprowadzania autobusów elektrycznych do eksploatacji w przedsiębiorstwach komunikacji miejskiej - wybrane zagadnienia. Prace Naukowe Politechniki Warszawskiej **112**, 217–226 (2016)
15. Murawski, J., Szczepański, E.: Perspektywy dla rozwoju elektro mobilności w Polsce. Logistyka **4**, 2249–2258 (2014)
16. InfoBUS. http://infobus.pl/pkm-jaworzno-i-kolejny-przetarg-na-autobusy-elektryczne-za-moment_more_109962.html
17. European Parliament Think Tank. http://www.europarl.europa.eu/thinktank/en/document. html?reference=EPRS_BRI(2019)637895

Modelling as Support in the Management of Traffic in Transport Networks

The Model of Arrival Time Distribution Forming in the Dense Vehicle Flow

Vladimir Chebotarev[1], Boris Davydov[2], Kseniya Kablukova[1],
and Vadim Gopkalo[2(✉)]

[1] Computing Center, Far Eastern Branch of the Russian Academy of Sciences,
Khabarovsk, Russia
vladimir.ch@ccfebras.ru, kseniya0407@mail.ru
[2] Far Eastern State Transport University, Khabarovsk, Russia
dbi@rambler.ru, vng@yandex.ru

Abstract. Dense vehicle traffic suffers impact of random factors, resulting in arrival delays at destinations. In this paper, we propose a stochastic model of the arrival deviation formation, which adequately reflects the influence of a set of independent factors on the travel time at section. The synthesis procedure of a travel time bimodal distribution, which reflects the mechanism of these factors impact, is described here. The consequence of the cumulative effect is the deformation of the arrival deviation density function with a change in the type of skewness. The developed model is in agreement with the statistical data obtained from Russian railways. The results of the study confirm the hypothesis that the diversity of the forms of the arrival deviation density function is caused by the presence of heterogeneous influencing factors.

Keywords: Stochastic model · Bimodal probability density function · Arrival deviation · Deformation of the probability density function

1 Introduction

Vehicles in the process of their traffic are influenced by random factors. This leads to schedule deviations which in turn cause subsequent delays. This is especially pronounced in a dense flow, where delays multiply, causing disturbances in the traffic of many vehicles. To predict the consequences of this process, it is necessary to study the formation laws of the arrival times and deviations (output deviations) distributions on the basis of known statistical characteristics of operating times. The results obtained can be useful in solving the tasks of dispatch control.

It should be noted that mechanisms of mutual influencing the trains on the railway and the vehicles in heavy city traffic are identical. It is difficult to change the order of vehicles on the highway in an intensive traffic. Therefore, proposed model of secondary delays describes both types of traffic flows.

This article examines the properties of the output deviation formation model, which takes into account random departure delays of the vehicle (for example, train) and the randomness of the travel time duration along a given route. Analyzing the real statistics, we noticed that the travel time often has a bimodal distribution type.

E. Macioszek and G. Sierpiński (Eds.): Modern Traffic Engineering in the System Approach
to the Development of Traffic Networks, AISC 1083, pp. 69–78, 2020.
https://doi.org/10.1007/978-3-030-34069-8_6

In addition, it was observed that from the station to station, the histogram of the travel time visually changes the skewness type. It should be noted here that it is customary to investigate skewness only for unimodal distributions. However, in the considered case of bimodality, skewness is also visible. For example, a histogram or a corresponding density graph of a random travel time duration between stations A_1 and A_2 has a bimodal form and a branch to the right of the larger mode, longer than the branch to the left of the smaller mode. But the histogram or the corresponding density graph of a random travel time duration between stations, say A_4 and A_5, may already have an opposite property. These observations served as a motivation for research.

The main random events leading to deviations in all types of traffic are technical failures of infrastructure and rolling stock. Also, the movement is outraged by abnormal weather conditions (fog, snow, etc.). The influence of the human factor is significant, e.g., when carrying out repair works. In addition, random train delays are caused by following problems at stations: uneven flow of passengers (in urban and suburban traffic) and deviations in the processing technology of trains (at cargo terminals).

It has been suggested that the appearance of bimodality is due to the influence on the train travel time at least two random factors. Such factors, in particular, are the train interactions in a dense flow and the occurrence of difficulties in the station functioning. In order to confirm this assumption, a model of forming the arrival deviation distribution taking into account the scattering of departure time and disturbances of the movement process over a section, caused by a pair of independent random factors, has been developed.

On the problem of formation the arrival deviation distribution, there is sufficiently small number of papers. These are reviewed in Sect. 2. The deviation formation stochastic model, taking into account the above mentioned factors, is described in Sect. 3. Section 4 shows the results of processing real data on the commuter and freight train traffic. Data analysis suggests that the developed model adequately describes the deformation property of the output distribution being under influence a pair of dominant random factors.

2   Literature Review

The research of the delays effect on the transport systems functioning has been the subject of numerous studies. In general, all proposed delay propagation models are divided into deterministic and stochastic. Deterministic models do not take into account the uncertainties that often arise in reality, and consider various traffic indicators (travel and dwell times, headways, etc.) as deterministic values. Effective models of delay propagation were proposed in [1, 2], the authors of which presented deterministic approaches based on the application of graph theory. Using these models allows dynamically adjust the vehicle traffic schedule.

In stochastic modeling, the influence of random factors is taken into account. The main provisions of the stochastic modeling are described in [3–8]. The authors of the classic stochastic models [3], [6] represent the total actual travel time as a sum of the travel times for the block-sections. Each travel time obeys the exponential distribution law (with a shift). As a result, the output distribution of the travel time obtained by them has always unimodal right-skewed form. However, real statistics obtained in a number of studies show that this type of travel time distribution is not always observed. Bimodal distributions of travel times [9] or headways on highways [10] are observed quite often.

A number of researchers have analyzed statistical data on train delays (see, e.g., [11–13]). All of them are unanimous in the opinion that the duration of such deviations is subject to the exponential distribution law. However, an investigation of real statistics shows that, in practice, the distribution of arrival deviations has more complex character.

Many researchers, analyzing the experimental time histograms of the travel time, believe the lack of data is a cause of the bimodality presence. However, they make a mistake when they approximate a density function by a smooth unimodal function. So, the authors of paper [9] try to describe a bimodal histogram by unimodal densities such as log-normal, log-logistic, etc. We state that a correctly designed bimodal density function is in better agreement with experimental data and provides a basis for a more complete analysis of the influencing factors.

Existing stochastic delay propagation models do not allow analytically calculating the probability of arrival deviations for all types of traffic disturbances. In our article, we are trying to improve the stochastic model, which is used to predict delays, given a random nature of the departure time and the combination of influences that lead to a complex pattern of the travel time dispersion along the section.

3 Stochastic Model of Arrival Deviations Formation

Consider the following stochastic model of a pair of vehicles. Vehicles with numbers 1 and 2 depart one by one from a certain starting point of the route and move in the same direction. Planned trajectories their movements are the same, only the moments of departure are different. It is assumed that the condition of conservation between vehicles a minimum safe distance is fulfilled. When departure and in the process of movement vehicles are influenced by random factors that lead to scattering of departure times and travel times.

We consider the most common scenario of arrival deviation occurrence. A conflict situation between vehicles occurs on open tracks between two control points. The difference in vehicle speeds and the departure deviations are the reasons for delay. The solution of this problem will allow extend the developed methodology to a multitude of vehicle interaction acts in the flow.

To describe the movement of a vehicle with the number i $(i = 1, 2)$ we introduce the following notations:

- δ_i - is the departure deviation from the schedule,
- $F(t) := \mathrm{P}(\delta_i \leq t)$ - is the distribution function of the random variable δ_i; it is assumed that the variables δ_1 and δ_2 are equally distributed,
- $f(t)$ - is the density function of δ_1,
- ρ_i - the travel time between starting and ending points of the vehicles,

- $L(t) = P(\rho_1 < t)$, $l(t)$ - is the density function of ρ_1,
- d_i - the departure time of the i-th vehicle according to schedule,
- a_i - the arrival time of the i-th vehicle according to schedule,
- ξ_i - the random deviation of the i-th vehicle from the arrival schedule; then $a_i + \xi_i$ is actual time of arrival of the i-th vehicle,
- $U_i(t) = P(\xi_i < t)$, $u_i(t)$ - is the density function of ξ_i,
- t_0 - the minimum permissible headway between vehicles,
- $s_0 = s_0(v) = v\,t_0$ - is the minimum safe distance between vehicles, corresponding to an arbitrary speed v.

Let us derive equality:

$$U_2(t) = \int_0^\infty F(t - d_1 + a_2 - t_0 - y)l(y)dy, \tag{1}$$

based on the arrival time formation scheme, shown in Fig. 1 (see [14]).

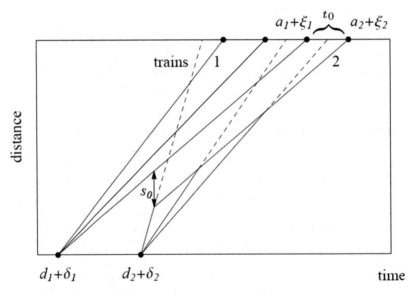

Fig. 1. The model scheme of the actual arrival time formation process

We can see from Fig. 1 that $a_1 + \xi_1 - (d_1 + \delta_1) = \rho_1$, hence $a_1 + \xi_1 = \rho_1 + \delta_1 + d_1$. Moreover, $a_2 + \xi_2 = a_1 + \xi_1 + t_0$. Thus, $\xi_2 = \rho_1 + \delta_1 + d_1 - a_2 + t_0$, and we obtain $U_2(t) = P(\rho_1 + \delta_1 < t - d_1 + a_2 - t_0)$. Assuming independency of ρ_1 and δ_1 we arrive at (1).

In turn, equality (1) implies the formula for the density function of ξ_2,

$$u_2(t) = \int_0^\infty f(t - d_1 + a_2 - t_0 - y)l(y)dy. \tag{2}$$

According to the collected statistics, the vehicle travel time often obeys some bimodal distribution law. Studies show that in the case of relatively free traffic, the graph of the density function of the time travel has a longer right branch. It is shown in [15] that such distribution is well approximated by the gamma distribution.

It can be assumed that the influence of several (in particular, two) factors on the train traffic is of a similar nature. Take the travel time density function $l(t)$ as a mixture of two gamma densities $l_1(t)$ and $l_2(t)$:

$$l(t) = l(t, p) = pl_1(t) + (1 - p)l_2(t), \tag{3}$$

where

$$l_1(t) = I(t > b_1)\frac{e^{-(t-b_1)/\beta}(t - b_1)^{\alpha-1}}{\Gamma(\alpha)\beta^\alpha}, \quad l_2(t) = l_1(t - b_2), \quad \alpha > 0, \quad \beta > 0. \tag{4}$$

The parameters b_1, $b_2 > 0$ set the shift of the initial gamma densities $l_1(t)$ and $l_2(t)$ along the horizontal axis to the right, because travel time is a strictly positive random variable. In this case, it is necessary to ensure that the support of the density function $l(t)$ lies only on the positive semi-axis, and it is significantly separated from zero.

Bimodality of the density $l(t)$ is determined by the influence of two random destabilizing factors. To simulate the degree of influence of each of them, we use the weighted parameter p, $0 \leq p \leq 1$. The more p, the stronger the first factor influences and the weaker the influence of the second factor, and conversely.

Let us set the actual form and parameters of the input distributions $f(t)$ and $l(t)$ in (2). Suppose that the departure deviation δ_1 obeys the exponential distribution law with a density function

$$f(t) = I(t > 0)\lambda e^{-\lambda t}, \quad \lambda > 0. \tag{5}$$

Let $\lambda = 0.3$. Such a value of the parameter λ corresponds to reality, because in this case, the mean departure deviation $E\delta_1 = 1/\lambda \approx 3.3$ min.

Next, we put $\alpha = 16$, $\beta = 0.25$, $b_1 = 6$, $b_2 = 3$. Denote by $\bar{\rho}_1$ and $\bar{\rho}_2$ the random variables with the densities $l_1(t)$ and $l_2(t)$ respectively.

Using the properties of gamma-distribution we obtain $E\bar{\rho}_1 = \alpha\beta + b_1 = 10$ min, $E\bar{\rho}_2 = \alpha\beta + b_1 + b_2 = 13$ min. The standard deviation for both random variables is equal to 1: $\sqrt{D\bar{\rho}_1} = \sqrt{\alpha\beta^2} = 1$ min.

The following parameters are determined by the normative schedule:

$$d_1 = 0\,\text{min}, \quad a_2 = 15\,\text{min}, \quad t_0 = 3\,\text{min}. \tag{6}$$

By varying the parameter p, let us see the behavior of the density function $l(t) = l(t, p)$ from (3), and its impact to the corresponding density $u_2(t)$, defined by equality (2). The results of the calculation are collected in Table 1.

Table 1. Behavior of the density functions $l(t) = l(t, p)$ and $u_2(t)$ for different values of the weighted parameter p.

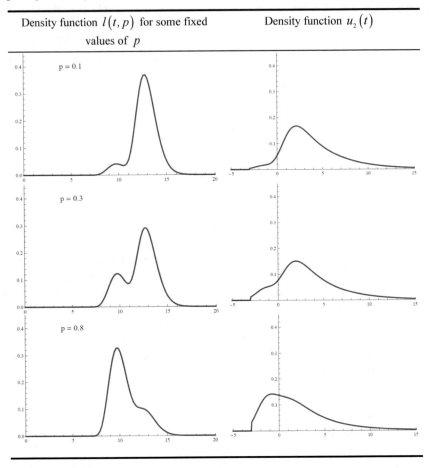

Density function $l(t, p)$ for some fixed values of p	Density function $u_2(t)$

When the parameter p is close to zero (see Table 1, $p = 0.1$), the first factor has a negligible effect on the travel time, the influence of the second factor prevails. Due to its impact, the mean travel time increases, which is accompanied by a higher right-hand vertex of the density graph $l(t, p)$ compared to the left one. And, as a result, the advance of the schedule is practically not observed: the density graph $u_2(t)$ is concentrated on the positive semi-axis.

As the parameter p increases, the degree of each factor impact changes: the mean travel time gradually decreases, the bimodal density $l(t, p)$ (see Table 1, $p = 0.3$) is changed by the unimodal one (see Table 1, $p = 0.8$). The deformation of the density $l(t, p)$ entails the deformation of the density $u_2(t)$. This leads to increase of the probability of advancing the schedule, namely the support of the density function $u_2(t)$ "captures" a segment of the negative semi-axis. Thus, the choice of the parameter p allows adjusting the degree of each factor influence depending on the specific situation.

The following patterns are observed. The smaller a value of the mean travel time, the smaller the mean of the delay. In addition, in this case negative arrival deviations are more often observed that is an advance of the schedule.

4   Validation of the Proposed Model by Using Statistical Data

This section is devoted to the analysis of research results by means of real statistical data of travel time and arrival deviations. The observed features are discussed taking into account the results of a theoretical analysis conducted in Sect. 3.

We explore the statistics collected for the trains of the suburban radial railway line Moscow-Tver. It turned out that bimodal travel time densities are not uncommon, but appear quite often. Table 2 contains the examples of histograms plotted from samples of travel time and arrival deviations distributions in the morning and evening rush hours. In Table 2 travel time and arrival deviations are measured in minutes (horizontal axis).

Note that for the convenience of displaying in travel time histograms, the vertical axis intersects the horizontal one not at the origin, but at the point (18, 0) or (5, 0).

The given examples show that the appearance of a bimodal travel time distribution leads to a rise in the left branch of the arrival deviation density, that is, it creates a left-sided skewness. If the influence of random factors on the train traffic is comparable, then it leads to the presence of two modes in the output density function. Thus, the nature of the histograms constructed on the basis of experimental data agrees well with the result of theoretical constructions, which are given in Sect. 3 of this article.

The histograms of the travel time given in Table 2 are fairly well approximated by a mixture of gamma distributions (3) with an approximately same degree of each factor influence on the travel time. In Table 1 this variant corresponds to the figures, which illustrate the densities of the considered random variables in the case when the weighted parameter p is equal to 0.3. For example, the following calculated parameters of the resulting arrival deviation density $u_2(t)$ characterize morning rush hours:

$$p = 0.3, \quad \alpha = 16, \quad \beta = 0.25, \quad b_1 = 17.5, \quad b_2 = 2.5, \quad \lambda = 1. \tag{7}$$

Moreover, the normative parameters for a given track are the following:

$$d_1 = 0\,\text{min}, \quad a_2 = 27\,\text{min}, \quad t_0 = 3\,\text{min}. \tag{8}$$

Figure 2 depicts a histogram of arrival deviations for morning rush hours from Table 2 with approximating density curve $u_2(t)$ with the parameters from (7), (8). Parameters correspond to the considered scenario. From Fig. 2 it can be seen that the theoretical density $u_2(t)$ from (2) is consistent with the experimental data. It should be noted that the degree of agreement can be increased by using the "cutting" of the density function $u_2(t)$ into the finite segment (the histogram support).

Table 2. Histograms of the travel time and arrival deviations distribution.

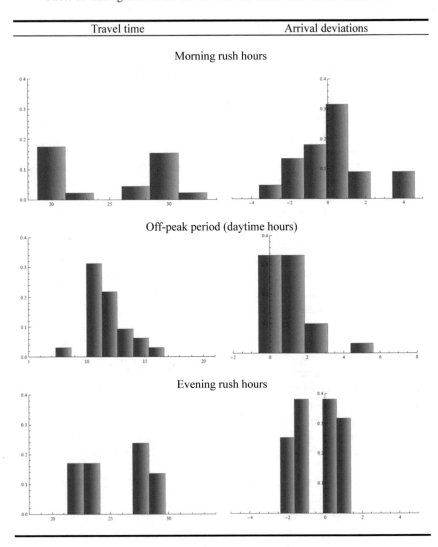

Results similar to passenger train traffic were obtained when considering statistical data on the freight train traffic through Trans-Siberian main line. The analysis shows that during periods of increase the train flow intensity, the number of sections in which a bimodal distribution of travel time is observed reaches 20%. Accordingly, the shape of the arrival deviation density curve becomes symmetrical or has left-sided skewness.

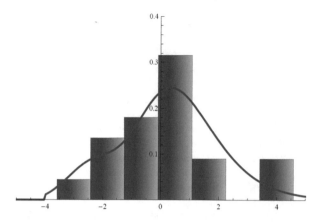

Fig. 2. Agreement of $u_2(t)$ with statistical data

5 Conclusions and Future Research

The stochastic model of the arrival deviation formation considered in this paper allows us to substantiate the assumption that the presence of independent influencing factors leads to a change in the shape and skewness of arrival deviation distribution. The flexibility of the model is that the resulting analytical formula allows using arbitrary theoretical distributions that adequately reflect the influence of these factors in various real-world scenarios.

The results obtained in the study reveal the mechanism of train arrival time distribution formation. This makes it possible to substantiate the method of vehicle traffic data statistical analysis, as well as to propose an adequate approximation of the arrival time distributions. This approach is useful both in daily traffic planning on a wide network of roads, and in the development of local solutions to compensate for deviations from the schedule in on-line mode.

Experimental data collected on Russian railways show that, in practice, bimodal travel time density is not uncommon and is observed quite often. Such nature of the travel time distribution also causes the deformation of the output arrival deviation distribution that is often observed in reality, that is, it acquires a negative skewness. The factor causing the bimodality of travel time density is often determined by mutual influence of the trains in flow due to the difference in their speeds. This fact allows us to build a more adequate train traffic model and increase the accuracy of delay prediction.

Subsequently, the authors plan to carry out an in-depth analysis of the arrival deviation formation process and apply the regularities identified in this research to a number of existing stochastic models of train traffic.

References

1. Muller-Hannemann, M., Schnee, M.: Efficient timetable information in the presence of delays. In: Ahuja, R.K., et al. (eds.) Robust and Online Large-Scale Optimization, vol. 5868, pp. 249–272 (2009)
2. Goverde, R.M.P.: A delay propagation algorithm for large-scale railway traffic networks. Transp. Res. Part C **18**, 269–287 (2010)
3. Carey, M., Kwieciński, A.: Stochastic approximation to the effects of headways in knock-on delays of trains. Transp. Res. Part B **28**, 251–267 (1994)
4. Carey, M., Kwieciński, A.: Properties of expected costs and performance measures in stochastic models of scheduled transport. Eur. J. Oper. Res. **83**, 182–199 (1995)
5. Yuan, J.: Stochastic modeling of train delays and delay propagation in stations. Ph.D. thesis, Technische Universiteit Delft, Delft (2006)
6. Meester, L.E., Muns, S.: Stochastic delay propagation in railway networks and phase-type distributions. Transp. Res. Part B **41**, 218–230 (2007)
7. Berger, A., Gebhardt, A., Muller-Hannemann, M., Ostrowski, M.: Stochastic delay prediction in large train networks. In: Proceedings of the 11th Workshop on Algorithmic Approaches for Transportation Modelling, Optimization, and Systems (ATMOS 2011), pp. 100–111. Dagstuhl Publishing, Germany (2011)
8. Buker, T., Seybold, B.: Stochastic modelling of delay propagation in large networks. J. Rail Transp. Plan. Manag. **2**(12), 34–50 (2012)
9. Lessan, J., Fu, L., Wen, C., Huang, P., Jiang, C.: Stochastic model of train running time and arrival delay: a case study of Wuhan-Guangzhou high-speed rail. J. Transp. Res. Board **2672** (10), 1–9 (2018)
10. Monamy, T., Haj-Salem, H., Lebacque, J.P.: Experimental analysis of trajectories for the modeling of capacity drop. Procedia Soc. Behav. Sci. **20**, 445–454 (2011)
11. Yuan, J.: Statistical analysis of train delays at The Hague HS. In: Hansen, I.A. (ed.) Train Delays at Stations and Network Stability. Trail, Delft (2001)
12. Yuan, J., Goverde, R.M.P., Hansen, I.A.: Propagation of train delays in stations. In: Allan, J., Hill, R.J., Brebbia, C.A., Sciutto, G., Sone, S. (eds.) Computers in Railways VIII, pp. 975–984. WIT Press, Southampton (2002)
13. Aleksandrova, N.B.: Distribution of the delay duration due to rejection by stations. In: Regional Scientific Conference: Siberian and Far East Universities for Transsiberian, Novosibirsk, pp. 20–21 (2002)
14. Chebotarev, V., Davydov, B., Kablukova, K.: Random delays forming in the dense train flow. WIT Trans. Built Environ. **181**, 435–445 (2018)
15. Chebotarev, V., Davydov, B., Kablukova, K.: Probabilistic Model of Delay Propagation along the Train Flow. https://www.intechopen.com/books/probabilistic-modeling-in-system-engineering/probabilistic-model-of-delay-propagation-along-the-train-flow

Methodological Bases of Modeling and Optimization of Transport Processes in the Interaction of Railways and Maritime Transport

Oleg Chislov[1]([☒]), Vyacheslav Zadorozhniy[1]([☒]), Dmitry Lomash[1],
Evgenia Chebotareva[1], Irina Solop[1], and Taras Bogachev[2]

[1] Rostov State Transport University, Rostov-on-Don, Russia
o_chislov@mail.ru, zadorozniy91@mail.ru,
lomash@mail.ru, abrosimova@yandex.ru,
bhbirfll22@yandex.ru
[2] Rostov State University of Economics, Rostov-on-Don, Russia
bogachev73@yandex.ru

Abstract. The analysis of the problems in ensuring railway traffic to the seaports is carried out in the given research. The ways of the development of planning system of railway transportation based on logistics and information technologies are proposed. Methodological approaches to the development of the program software to manage export transportation in rail and maritime traffic are developed. Simulation principles of interaction procedures of different transport modes are stated, models and algorithms of adaptive approach of trains to the port stations are developed. Spheres of simulation modeling method application and method of multi-agent optimization Ant Colony System (ACS) for solving transportation tasks in the system "Railway - Seaport" are considered. The presented mathematical model provides ample opportunities to solve transport regulation issues. The theory of higher order algebraic curves is used to create a territorial model of an oligopolistic freight market.

Keywords: Modeling · Simulation modeling · Optimization · Software package · Freight traffic · Railways · Sea ports · Transport hub · Planning · Organizing · Management · Operator company · Oligopolistic market · Influence areas

1 Introduction

Under expansion of the foreign trade relations the development of new methods in solving the problem of optimizing the interaction of several transport modes is required. The analysis of theoretical positions and practices show that the interaction of various transport modes is based on a number of economic, technological, technical, organizational and managerial conditions such as:

- ensuring the identity of plans for the freight transportation in mixed traffic,

E. Macioszek and G. Sierpiński (Eds.): Modern Traffic Engineering in the System Approach to the Development of Traffic Networks, AISC 1083, pp. 79–89, 2020.
https://doi.org/10.1007/978-3-030-34069-8_7

- conformity of structural and power unification of transport elements of various modes of transport, required throughput and processing capacity of components and devices of the transport infrastructure,
- ensuring the required capacity of rehandling facilities and storage,
- availability of appropriate shunting means and devices,
- development of rational technology of railway transport divisions' operation and others.

As a result of mismatch in any of specified parameters or uncoordinated actions of transportation participants there are losses at the junctions of the interacting modes of transport. A negative image of the transport system that does not attract additional volumes of the international freight traffic is formed.

The main area of the research is the organization of rail transportation to seaports. In recent years volumes of transportation to ports by railways is constantly increasing. If we consider seaports of Russia, the main port potential is concentrated in 44 seaports of 62 Russian ports. The largest volumes of freight delivery fall on the railway-maritime traffic and at the same time growth of quantitative indices of operation of the adjacent modes of transport is noted (Table 1).

Table 1. Operation indicators of railway and maritime transport of the Russian Federation [1].

Indicators	Years			
	2005	2010	2015	2017
Railway transport				
The volume of freight traffic to the sea ports of Russia, total, mln. tons	178.0	219.5	305.3	333.9
Including export	148.3	196.8	264.1	293.4
The intensity of freight traffic per 1 km of public railway lines length, mln. t-km	21.8	23.5	26.7	28.8
Freight turnover, bln. t-km	1858	2011	2306	2493
Maritime transport				
Transported cargo, mln. tons	26.0	37.0	18.8	24.6
Cargo turnover, bln t-km	60.3	100.3	41.7	45.9
Overloaded cargo at berths in seaports total, mln. tons	135.8	526.1	676.7	786.4
Including export	106.9	404.2	539.1	605.8

In many other countries of the world rail transportation also plays a significant role in the transport system. The analysis of railway operation servicing ports showed that there are significant difficulties in organizing the passage and processing of an export car traffic volume. Uncoordinated actions of participants of the transport process also lead to a decrease in quality of transport services, failure within delivery periods, and an increase in transportation costs. Despite decades of rich experience in the development of the management system of mixed railway and maritime transportation the main problems of transport modes interaction remain unchanged.

The sphere of research includes the theory of product distribution management, the theory and practice of complex systems modeling, information support of goods transportation and irregularity of transport production processes (researches by Forrester, Shannon, Schreiber, Pottgoff and etc. [2–5].

The methodical basis of this research is methods of the systemic-functional analysis, mathematical and simulation modeling, expert procedures, etc. To identify problems of transport modes interaction statistical reporting of railways and maritime transport operation is used. The authors have defined classes of tasks taking into account different areas and methods of modeling (Table 2).

The methodology of the solution of the transport tasks presented in Table 2 is given below.

Table 2. Classes of assignments and used optimization methods in the interaction of rail and maritime transport.

No.	Assignment	Type and modeling method	Object of modeling
1.	Optimization of freight traffic at transport polygon	Economical and geographical model	Transport polygon
2.	Optimization of transport and logistics chain	Queueing system	Transport and logistics chain, transport hub, seaport
3.	Optimization of transport hub operation, train approach to the transport hub, adjustment of car shipment schedule to seaports	Simulation modeling	Transport hub, cargo fronts, seaport

2   Economic-Geographical Model of Delimitation of the Freight Market on the Polygon

Consider the sequence of building an economic - geographical model, giving a territorial picture of the optimization of the delimitation of freight traffic on the transport services market in a given region.

At the first stage, on the basis of graph theory and a system of analytical calculations, a Geometric Euclidean Model (GEM) of a road test site is formed, which is a cube-matrix (Fig. 1).

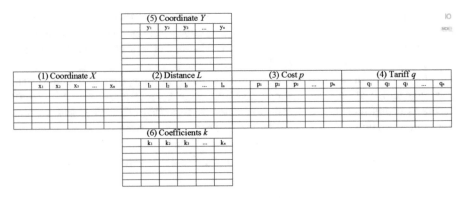

Fig. 1. Expanded GEM cube

Next, a statement of the values of freight transportation of the i-th carrier company in the rolling stock is formed (Table 3).

Table 3. Statement of values of freight traffic.

	Station loading	Port (destination station)					
		T		H		...	
		L, km	C, ruble.	L, km	C, ruble.	L, km	C, ruble.
Company operator	A	L_{A-T}	C_{A-T}	L_{A-H}	C_{A-H}	...	...
	B	L_{B-T}	C_{B-T}	L_{B-H}	C_{B-H}	...	...
	C	L_{C-T}	C_{C-T}	L_{C-H}	C_{C-H}	...	...
	G	L_{G-T}	C_{G-T}	L_{G-H}	C_{G-H}	...	...
	...	...	...	...	...	...	...
	...	...	...	...	...	...	...
	N	L_{N-T}	C_{N-T}	L_{N-H}	C_{N-H}	...	...

As a result of the least squares processing of numerical data from Table 3, for each i-station the corresponding analytical expression of the cost of transportation is found c, ruble from distance l, km (Table 5).

These distances and values are ranked from minimum to maximum (Table 4).

Table 4. Ranking distance and cost data.

Port (Destination station)	Station loading					
	A		B		...	
	L	C	L	C	L	C
T	L_{min}	C_{min}	L_{min}	C_{min}	...	...
H	L_{min+1}	C_{min+1}	L_{min+1}	C_{min+1}	...	...
...	...	...	...	...	...	...
...	L_{max-2}	C_{max-2}	L_{max-2}	C_{max-2}	...	...
...	L_{max-1}	C_{max-1}	L_{max-1}	C_{max-1}	...	...
N	L_{max}	C_{max}	L_{max}	C_{max}	...	...

Table 5. Dependence of the cost of transportation on the distance.

Station A	Station B	Station C	...
$C_A = p_k + q_k \cdot l$	$C_B = p_k + q_k \cdot l$	$C_C = p_k + q_k \cdot l$	...

Depending on the values of the coefficients q_k for the length l_k of the «influence area» route of the k-stations, they are delimited by either Descartes ovals (in particular, Pascal's limaçon) or curves of the 2nd order (hyperbola branches or a circle).

The geometrical idea of the approach used by the authors goes back to [6]. It received a significant development analytically in [7]. As a result, a geometrical Euclidean model (GEM) of the freight transportation market was proposed. In this model, the real situation is «idealized» in the sense that all transport routes are assumed to be the straight. Thus, their lengths are Euclidean distances between the beginnings and the ends of the routes. We give a brief description of the theoretical basis of GEM. For comparison, let us cite a paper [8] in which other methods are used and a geometric model is build for the probabilistic port hinterland based on intermodal network flows jointly using discrete choice analysis and geographical information of shippers.

We first consider the case of a duopoly. The delimitation between the «influence areas» of the two subjects of the transportation process (duopolists) will be determined by the cost of transporting goods from the duopolist to the destination. The destination can be, in principle, at any point of the plane. Let L be the distance between the locations of duopolists, $p_i > 0$ and $q_i > 0$ are the costs of initial-final operations and the cost of movement operations per 1 km of the route, spent (for example, on one wagon) by the i duopolist ($i = 1, 2$). Without loss of generality, we will assume that the 1st and 2nd duopolists are located respectively at the points of the plane $O(0,0)$ and $A(L,0)$. Then the curve delimiting their «influence areas» is written by the equation:

$$p_1 + q_1 \sqrt{x^2 + y^2} = p_2 + q_2 \sqrt{(x - L)^2 + y^2}. \tag{1}$$

Equation (1) defines an algebraic curve on a plane, which generally has a fourth order.

Since the corresponding analytical calculations are cumbersome, it is convenient to perform the transformations of Eq. (1) in the computer math environment.

Let be $p = p_2 - p_1$, let's pretend that $p \geq 0$. The result will be the equation:

$$\begin{aligned}
&(q_1^2 - q_2^2)^2 x^4 + 4q_2^2 L(q_1^2 - q_2^2)x^3 + 2(q_1^2 - q_2^2)^2 x^2 y^2 - (4p^2 q_2^2 + 2(p^2 + q_2^2 L^2))(q_1^2 - q_2^2) \\
&- 4q_2^2 L^2)x^2 + 4q_2^2 L(q_1^2 - q_2^2)xy^2 + (8p^2 q_2^2 L - 4(p^2 + q_2^2 L^2)q_2^2 L)x + (q_1^2 - q_2^2)^2 y^4 \\
&- (4p^2 q_2^2 + 2(p^2 + q_2^2 L^2)(q_1^2 - q_2^2))y^2 - 4p^2 q_2^2 L^2 + (p^2 + q_2^2 L^2)^2 = 0.
\end{aligned} \tag{2}$$

We give some geometric explanations about the curve given by the implicit Eq. (2).

Let $q_1 > q_2$ and $p_2 > p_1$ (that is $p > 0$). Then the curve under consideration has the fourth order and is actually the Descartes oval. If $q_1 = q_2$, then the order of the curve is reduced by at least two orders of magnitude. In this case, depending on the relationship among the values of the parameters p, L and q_2 the shape of the curve can vary significantly. Namely, a curve is a right branch of a hyperbola (that is, a curve of second order) if $0 < p < Lq_2$. It becomes a straight line ray if $p = Lq_2$. Finally, if $p > Lq_2$, then the curve is an empty set.

Note that the order of the curve defined by Eq. (2) can also decrease significantly in cases when $q_1 > q_2$. This happens if and only if equality $p_2 = p_1$ (that is $p = 0$) holds. In such cases, the specified curve is a circle (a second order curve).

Let us give the comments on the «influence areas» received by participants in a duopoly.

The software complex, which realizes for the practical use the original economic-geographic model constructed in the environment of the analytical calculation system, has been created on the basis of the Java software platform (Fig. 2).

The economic-geographical model of the rational distribution of railway traffic volumes at the site of the port railway is presented in a user-friendly interactive form in the software complex. At the same time, it is possible to edit timely economic indicators in operational activities for all the participants of the transportation process.

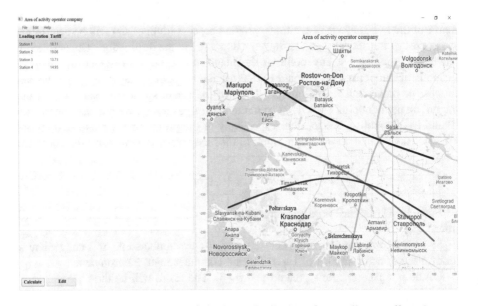

Fig. 2. The software complex of the rational distribution of port railway traffic volume

3 Application of the Queueing Theory in Optimization of Transport and Logistics Chain of Export Traffic Volumes Delivery to Seaports

In railway station and junction complexes, unloading areas and ports there are modules that can be described by a wide range of queueing systems in case of unequal requests and modules are specified by certain discipline of priorities. Delivery of freight from point A to point B is carried out by passing through several interrelated stages (Fig. 3) which include: freight clearance procedures, its processing at station yards, storage points and ports, transportation on railway sections.

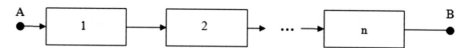

Fig. 3. Transport and logistics chain as a queueing system

The technology of analysis based on the concepts of queueing theory is well known [9, 10]. Each block of Fig. 3 is a separate service channel.

The analysis considers the input and output flows, the system structure. The parameters of the channels are calculated: the average service time of the request, service waiting time, queue length, etc. Comparing these parameters in the logistics chain, one can determine its weak link, the reserves to increase carrying capacity. In the theory the methods for the simplest flow requests (input and output) are well developed. They must be stationary, ordinary, have the property of the absence of aftereffect. Such requirements for traditional traffic flows are obviously not always met. This constrains the application of the method. Therefore to solve the problems of strategic and operational levels of management simulation modeling is widely used.

4   Methodology of Simulation Modeling of Processes in the Transport Hub and Trains Approach to Seaports

To model the operation of complex systems, which include the transport network and its elements, it is advisable to use simulation modeling. For several years the authors have been carrying out the research of simulation modeling application for solving transport problems in the organization of rail and maritime transportation and an approach to choosing simulation modeling tools based on the hierarchy analysis method has been developed. The transport hub Novorossiysk as a complex system handling the largest seaport is presented as a multi-level structure consisting of a number of interacting elements combined into subsystems of various levels [11]. At decomposition of the process of interaction of the station Novorossiysk and port it was established that the performance time of technological operations can be described by means of probability distributions. On the basis of simulation modeling the impact of various external factors on functioning and stability of model is estimated. For example, during the simulation, the influence of the number of days of non-acceptance of trains by the station due to adverse weather conditions on the value of car turnover was estimated. Additional tests made it possible to evaluate the efficiency of trains' approach to the hub with different freight nomenclatures using multi-agent optimization methods. The essence of the additional tests was in comparing the values of the car turnover at closing of the port for one, two and three days. The results of 520 runs of the model are shown in Fig. 4.

The obtained dependences confirmed the fact that the greatest effect, taking into account the application of the multi-agent optimization method for determining the sequence of train access, is achieved with non-failure operation of the port. With an increase in the number of non-acceptance days of trains by the station the effect decreases. For example, if the port does not accept cargo within three days due to unfavorable conditions, the advantages of application of optimization methods are minimal.

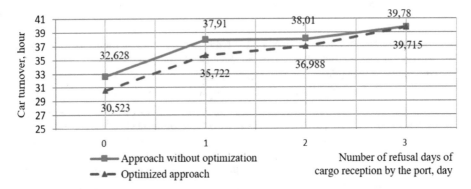

Fig. 4. The change of the value of the car turnover at different strategies of trains' approach to the port

To simulate trains' approach to the port, a discrete-event apparatus is proposed. The transition graph between states for system transactions is shown in Fig. 5. The first step includes information input on loading at the stations of freight origin. A simulation model is being developed taking into account statistical travel time of trains to the port and other data. After entering the initial data the model run is made that allows to simulate the movement of cargo and its unloading at the port (Fig. 6).

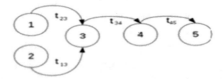

1 – Freight origin at the station (1)
2 – Freight origin at the station (2)
3 – Freight is on the polygon of the port road
4 – Freight is in the zone of responsibility
    of the port station
5 – Freight is unloaded on the vessel

Fig. 5. Transition graph between states

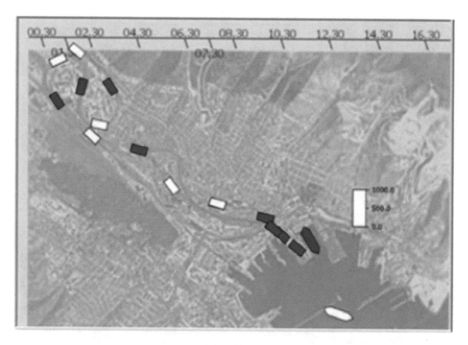

Fig. 6. Simulation model of the polygon of freight approach

5 Rescheduling the Shipment of Cars Using the Ant Colony System Algorithm

To ensure the smooth operation of the railways on unloading and to exclude the appearance of trains idle waiting for operations on the road polygon or at the port station, it is necessary to synchronize the loading and unloading of freight at loading stations and at transshipment points. This task in the researches on the planning of rail and maritime transportation has not been solved yet. In order to ensure regular operation, the authors have developed a software package that solves the problem of optimizing the planning and adjusting the dates and volumes of loading. In the software package the adjustment block contains an optimization model for selecting the adjustment station and allows to determine those stations at which it is recommended to reduce the loading volume on a specific day. The results of the optimization model are forwarded to the input of the simulation model. Another run helps to build up a forecast model of unloading at the port station (Fig. 7). To solve this problem a heuristic optimization method based on the multi-agent optimization method Ant Colony System (ACS) proposed by Dorigo, Maniezzo and Colorni [12] is used in the software.

Railways and seaports have been working together for many years in search of ways to optimize the transportation process. At present the work is being carried out to implement the Road Information and Logistics System (RILS) for the North Caucasus Railways. RILS is the object of the Digital Railways with elements of cyber-physical systems, intellectualization of decision-making procedures, blockchain, naj-technology. The subtasks of the RILS software complex are shown in Fig. 8.

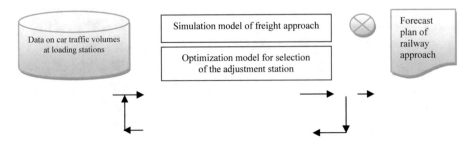

Fig. 7. Program complex of optimal freight approach to the port road

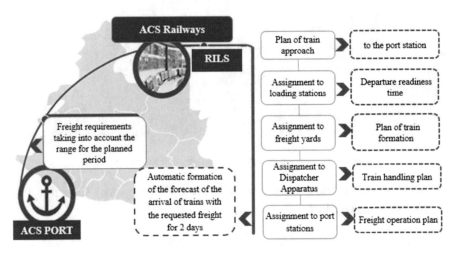

Fig. 8. Subtasks of RILC program complex

6   Conclusion

The «Railway - Seaport» system is complex, therefore this paper focuses on managing the loading process and approach of freight taking into account ports uploading capacities and using the simulation modeling. For example, the interaction simulation of the port station and the port was carried out to assess the influence of the factor of application of control actions on the efficiency of the port station operation (car turn-over) with and without optimizing methods. The application of software solutions proposed by the authors based on simulation modeling will allow to reduce unproductive idle time of running vehicles while preserving the existing volume of traffic and to speed up cars movement to seaports.

References

1. Transport in Russia 2018: Federal State Statistics Service, Moscow (2018)
2. Forrester, J.W.: World Dynamics. AST, Moscow (2003)
3. Shannon, R.E.: Systems Simulation. University of Alabama in Huntsville, Huntsville (1975)
4. Schreiber, T.G.: Modeling on GPSS. Mashinostroyeniye, Moscow (1980)
5. Pottgoff, G.: Study of Traffic Flows, Moscow (1975)
6. Krynsky, H.E.: Mathematics for Economists: Statistics, Moscow (1970)
7. Chislov, O., Bogachev, V., Zadorozhniy, V., Bogachev, T.: Economic-geographical method delimiting wagon flows in the region considered: model and algorithm. Transp. Probl. **13**(2), 39–48 (2018)
8. Wang, X., Meng, Q., Miao, L.: Delimiting port hinterlands based on intermodal network flows: Model and algorithm. Transp. Res. Part E Logist. Transp. Rev. **88**, 32–51 (2016)
9. Lyabakh, N.N., Butakova, M.A.: Queueing Systems: Theory Development, Methodology of Modeling and Synthesis. RSTU, Rostov-on-Don (2004)
10. Takha, Kh.A.: Introduction Into Operations Research, 7th edn., Moscow (2005)
11. Chernyaev, A.G., Chebotaryeva, E.A., Lomash, D.A.: The Development of the Export Transportation Management System in Railway and Maritime Traffic on the Basis of Logistic and Information Technologies. RSTU, Rostov-on-Don (2013)
12. Dorigo, M., Gambardella, L.M.: Ant colony system: a cooperative learning approach to the traveling salesman problem. IEEE Trans. Evol. Comput. **1**, 53–66 (1997)

Web Planning Tool for Deliveries by Cargo Bicycles in Kraków Old Town

Vitalii Naumov[1](✉), Hanna Vasiutina[2], and Jakub Starczewski[1](✉)

[1] Faculty of Civil Engineering,
Cracow University of Technology, Kraków, Poland
vnaumov@pk.edu.pl, starczewski.jakub@onet.pl
[2] Faculty of Physics, Astronomy and Applied Computer Science,
Jagiellonian University, Kraków, Poland
hanna.vasiutina@student.uj.edu.pl

Abstract. The paper proposes a mathematical model and implemented software for a freight delivery system, where cargo bicycles are used as a mean of transport. The proposed mathematical model uses a graph representation to describe a transport network, where consignees are associated with the vertices, while the graph edges reflect the sections of the road net. Requests for a freight delivery are characterized with a set of numerical parameters, such as: the consignment size, the package dimensions, the time interval between requests from clients and the delivery distance. The mathematical model was used in the web planning tool for deliveries by cargo bikes in the Kraków Old Town. In the paper, the authors briefly describe some features of the developed web application: the procedures of making orders by freight owners, creation of the delivery routes for carriers and other planning and management operations.

Keywords: Bicycle · Routes planning · Decision support software

1 Introduction

Cycle logistics is a new and dynamically developing area of science. Results of its fast growth are clearly noticeable: huge delivery companies use cargo bikes for distribution of goods to the customers, also cargo bikes are widely popular in daily people's routine (for shopping and child transport). However, development of this area has to be correlated with development of respective practical tools. According to the reviewed literature, cargo bikes transport systems have many limitations, therefore, to effectively use them in a commercial transportation some special conditions has to be set [1]. First of all, it is a distance of carriage, which usually (in 90%) lies within the range of up to 75 km [2]. Because of that, to use cargo bikes in practice it is necessary to adapt a loading point (one or more), which allows to reduce the delivery distance for the bikes. There are also some problems with the loading points: where is the optimal place of their location in the city area and which method of calculation is better to use [3]. As the sources show, the method of gravity center, heuristic methods or some others (for example, based on analyses of transport work) can be used [3, 4]. Secondly, only some types of freights can be carried by cargo bicycles. There is still a need for special

© Springer Nature Switzerland AG 2020
E. Macioszek and G. Sierpiński (Eds.): Modern Traffic Engineering in the System Approach
to the Development of Traffic Networks, AISC 1083, pp. 90–98, 2020.
https://doi.org/10.1007/978-3-030-34069-8_8

researches on the transport susceptibility of cargo, but practical usage shows that the small, light (for example: courier-size) packages, that must be distributed in the old city centers, can be a suitable freight for delivering by bike. Such issue is related to an increase in importance of the last mile delivery problem as a highly cost-intensive stage in transportation process [5]. Therefore, cycle logistics is still looking for an answer to the question: how to effectively use cargo bikes in distribution process. On this basis, a project named City Changer Cargo Bike (CCCB) was started under the Horizon2020 program. It has the following goals [6]:

- raise awareness among the relevant interested parties: public, private and commercial sector,
- make use of the innovative tools and transfer between forerunner and follower cities (peer-to-peer exchange),
- establish favorable framework conditions for cargo bike usage,
- achieve wide deployment and transferability through forerunner cities, follower cities (within the consortium) and external follower cities,
- reduce traffic jams, emissions; increase safety; increase and improve public space usage.

There are 20 partner cities from all over Europe involved in the project and its value is: 3 920 712,50 EUR. One of the main tasks of the CCCB project is the support of the development of IT services, that ensure the effective functioning of the city's transport system. As a part of this task, the web application: "Cargo Bikes" that supports users of the cargo bike transport system was created. This study describes the main functionality of the developed tool.

2 Mathematical Model of Software

The described application is a website, which contains tools for supporting deliveries by cargo bikes in the old city district of Kraków. The service is based on the earlier analysis of subject [6] and implements the mathematical model described below.

2.1 Transport Network Model

Modeling of transport networks is based on the use of mathematical structures, that depend on a graph models. As a result, nodes of the transport network represent the customers (for which the cargo is to be delivered), while the links reflect the relevant sections of the road network - connections between the customers. The model of the transport network Ω can be described as:

$$\Omega = \langle \{\eta_i\}, \{\lambda_j\} \rangle, \forall \eta_i \in N, \forall \lambda_j \in \Lambda, \tag{1}$$

where:

η_i -is a node, which is an element of the set N of all nodes of the transport network,
λ_j -is a link, which is an element of the set Λ of all links of the transport network.

The basic characteristics of the nodes are their geographical coordinates (for example, GPS coordinates) and lists with input and output links:

$$\eta = \langle x, y, \lambda_{in}, \lambda_{out} \rangle, \tag{2}$$

where:
x and y - are geographical coordinates of the node,
λ_{in} and λ_{out} - are sets with input and output links for the node, $\lambda_{in} \subset \Lambda$, $\lambda_{out} \subset \Lambda$.

The main parameters of links are its weight and vertices (start and end):

$$\lambda = \langle w, \eta_{out}, \eta_{in} \rangle, \tag{3}$$

where:
w - is a weight of the link (for example, length, cost or time of travel),
η_{out} and η_{in} - are beginning and end node of the link, $\eta_{out} \in N$, $\eta_{in} \in N$.

2.2   Model of Demand for Goods Delivery

The basic variable, that defines the demand, is a request for a delivery of goods, which describes the customer's demand for services. Each request can be quantified with a set of numerical parameters, among which the most significant are the size of the cargo, its dimensions, time interval between requests and the delivery distance. In general, a demand model can be presented as an orderly (by the time) set of requests. The demand model D is described as follows:

$$D = \{\rho_1, \rho_2, \ldots, \rho_n\}, \tag{4}$$

where:
ρ_i - is the i-th request in a queue: $\rho_i \prec \rho_{i+1}$, where $t_i \leq t_{i+1}$, t_i is the moment of the i-th request appearance,
n - is the number of requests in a queue.

A single request for a cargo bike transport system can be described as:

$$\rho = \langle \eta_o, \eta_d, \zeta, \omega, \langle \theta_l, \theta_w, \theta_h \rangle \rangle, \tag{5}$$

where:
η_o and η_d - are the nodes, that define the location of the consignor and consignee, $\eta_o \in N$, $\eta_d \in N$,
ζ - is the time interval between the appearance moments of new request and the previous request [min.],
ω - is the weight of the request [kg],
$\theta_l, \theta_w, \theta_h$ - are dimensions of the cargo unit - its length, width and height [cm].

For a single request these parameters are deterministic, but for the queue of requests, they will be random variables. Because of that, the task of a demand modeling can be presented as a task of generation a queue of requests with random parameters.

$$D = \left\langle \Delta, \tilde{\zeta}, \tilde{\omega}, \left\langle \tilde{\theta}_l, \tilde{\theta}_w, \tilde{\theta}_h \right\rangle \right\rangle, \tag{6}$$

where:
Δ - is the trip matrix for all consignors and consignees from queue of requests,
$\tilde{x}$ - is a random variable, which characterizes some numerical demand parameter x.

3 Proposed Implemented Software

3.1 Technologies and Architecture of the Software

For the implementation of the "Cargo Bikes" web application that supports users of cargo bikes transport system in Kraków city center the Full Stack Java Script Development approach was used. This technology has many advantages and allows to use Java Script programming language in all layers of the client-server type of application, which is significantly simplifies the programming process. The front end part of the application has been implemented by using standard tools: HTML, CSS and Bootstrap, an open source toolkit, which ensures the responsiveness of the user interface. Moreover, the back end part is implemented with help of the Node.js technology and its modules: express, bcrypt, body-parser, mongoose and google-distance. The applications, which are implemented with Node.js technology and its express framework, do not require strict rules regarding to the project structure. However, described application is built according to the Model-View-Controller architecture template, which is probably one of the most popular ones. The MVC template consists of:

- model - represents the data structure, the format and the constraints with which it is stored. It maintains the application data and is a part of the database,
- view - is a content, that is presented to the user. It uses the Model and presents data in a form in which the user wants. The user may also be able to make changes to the data presented to him. View consists of static and dynamic pages, which are rendered or sent to the user when the user requests them,
- controller - receives and interprets user requests and then generates the response to the viewer. The user interacts with the view, which generates the appropriate request. This request is handled by the controller, which then will display the corresponding view along with the model data as a response.

This approach for creation an application structure is implemented in the described "Cargo Bikes" project, which has 'models', 'views' and 'routes' directories in its structure. The general tasks of those elements are: "models" provides manipulation on a data, "views" manages the user interface, and "routes" maintains request-response handling. Figure 1 shows an application structure.

Fig. 1. The structure of the application

The description of the elements of the application structure is given below:

- the "routes" folder contains files that determine how the application responds to the client's request to the specified URI end point and specific HTTP request methods (GET, POST, etc.) are handled. Files contained in this folder represent controller,
- the "views" folder contains template files, that implement user interface. File extension.ejs means that the EJS template engine is used to render the content to the user,
- the "models" folder has files, that implement the database collection schemas.

The main file, that runs the "Cargo Bikes" application, is *app.js*. The "app" object conventionally means the Express application and is created with the top-level *express* () function, that is exported by the Express module using the *require*() function. The app object has methods to: route HTTP requests, configure the middleware, render HTML views, register the template engines, etc.

3.2 User Interface

The application user interface consists of the main page, subpages for registering and logging in clients and subpages for managing orders and user profile data. During the registration, users are divided into two types: "cargo owner" and "carrier". Depending on selection of the user type, the appropriate information is displayed to the customer. The "cargo owner" has the ability to manage requests for the cargo delivery and the "carrier" gets optimized delivery routes. The "views" folder contains files that implement user interface. These are: home page (*index.ejs*), registration and login subpages (*register.ejs* and *login.ejs* respectively). After registration or login procedure user is redirected to the user profile page: *ownerpage.ejs* for cargo owners or *carrierpage.ejs* for carriers. The user profile pages have subpages: for adding new request or editing the existing one by the cargo owner (*newConsignment.ejs* and *editConsignment.ejs*), and for editing profile data for both types of users (*editUserProfile.ejs*). The home page contains the user's navigation menu, that moves user to the "consignments", "city map" and "contact" page sections or redirects to the registration or login subpages. Figure 2 shows a fragment of the main page that displays information about requests.

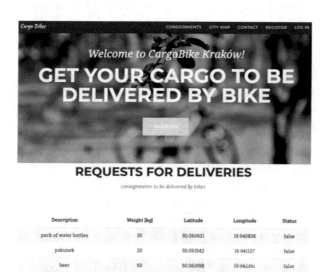

Fig. 2. A fragment of the main page, that displays the information about requests of the cargo owners

The application allows to render on a map a localization of the cargo delivery points, obtained from the application database. The view with the markers, that point to the requests on the city center map of Kraków, is presented in Fig. 3.

Moreover, the application consists of other sections and subpages, that are characterized below:

- *"Contact us"* section - allows user to quickly create and send an e-mail with a message,
- *"Register"* form - contains input fields that provide user data to the "users" collection of the database,

KRAKÓW CITY MAP

Fig. 3. Marker points on the city center map of Kraków

- *"Login"* form - identifies the registered user and (if the user enters correct data) redirects him to the profile page after the "Login" button is clicked. If the user enters incorrect data (name or password), error messages will be displayed,

- *"Cargo Owner"* page-consists of the sections that display personal data, own requests and a map with markers that point to the request locations. Clicking the "New consignment" button redirects owner to the page with form for adding a new request. "Delete" button deletes existing request. "Edit" button redirects owner to the form with editing request functionality. The "Logout" button redirects user to the home page,

- *"Carrier"* page is implemented for the "carrier" type of user. It is divided into subsections: personal data, possible delivery routes and a map with markers that point to the location of the delivery points and the loading point. Carrier can choose a delivery route by clicking the "Get Route" button, which updates the properties of the "droutes" collection in database. The route status and the status of all requests in this route will change to "true" and the carrier ID will be assigned to ID value. The selected route will be excluded from the list of possible routes to be performed by other users. Figure 4 shows a view of the "Carrier" page displayed on an iPad,

- *"Edit User Profile"* subpage implemented for every registered user to have the possibility to edit personal data or delete a profile. New data is retrieved from the form and by clicking the "Save" button is saved to the database. After the "Delete user" button is clicked, a modal window is opened, asking for the confirmation to delete the account.

The website functionality allows users to comprehensively plan and manage the cargo bicycles delivery process in the old city center of Krakow. The described application is a complete and practice IT tool for all user groups.

Fig. 4. The "Carrier" profile page displayed on an iPad

4   Conclusions

Cargo deliveries performed by bikes is a good response to the last mile delivery problem, however, there is still a need in practical tools to support this process. The proposed application is a comprehensive IT tool for implementing bicycle deliveries. Its construction and structure ensure efficient operation of the transportation market entities. Presented mathematical model and its software implementation allow to handle the requests and to generate delivery routes in a convenient manner for users. It is possible to expand the application with new modules, which gives possibility to integrate deliveries in the real-time mode and supervise them.

References

1. Naumov, V., Starczewski, J.: Choosing the localization of loading points for the cargo bicycles system in the Krakow old town. In: Kabashkin, I., Yatskiv (Jackiva), I., Prentkovskis, O. (eds.) Reliability and Statistics in Transportation and Communication. LNNS, vol. 68, pp. 353–362. Springer, Cham (2018)
2. Gruber, J., Kihm, A., Lenz, B.: A new vehicle for urban freight? An ex-ante evaluation of electric cargo bikes in courier services. Res. Transp. Bus. Manag. **11**, 53–62 (2014)
3. Iwan, S.: Zarządzanie dostawami ostatniego kilometra realizowanymi z wykorzystaniem rowerów towarowych. Prace Naukowe Uniwersytetu Ekonomicznego we Wrocławiu **383**, 867–880 (2015)

4. Naumov, V., Starczewski, J., Szarata, A.: Wybór lokalizacji punktu przeładunkowego na potrzeby rowerowego systemu dostaw ładunków. Prace Naukowe Politechniki Warszawskiej **120**, 309–318 (2018)
5. Chodak, G., Łęczek, J.: Problem ostatniej mili - wyniki badań sklepów internetowych i konsumentów. https://mpra.ub.uni-muenchen.de/68851/1/MPRA_paper_68851.pdf
6. CORDIS. https://cordis.europa.eu/project/rcn/215999/factsheet/en

Charging Station Distribution Model - The Concept of Using the Locations of Petrol Stations in the City

Marcin Staniek[✉] and Grzegorz Sierpiński

Faculty of Transport, Silesian University of Technology, Katowice, Poland
{marcin.staniek,grzegorz.sierpinski}@polsl.pl

Abstract. In cities of low level of electromobility, it is particularly important to plan possibly the most efficient distribution of first-established charging stations. Since it contributes to building trust to electric vehicles, locations of charging stations should correspond to the actual needs of users to promote electromobility and maximize its implementation effect. The article presents a decision-making support method that helps to determine locations of the first charging stations in a given area. The method is based on the assumption that charging stations are set up at existing petrol stations. The method has been applied for the area of the city of Sosnowiec.

Keywords: Charging station · Electromobility · Transport systems planning

1 Introduction

Transport generates certain challenges, in particular the congestion in city centers and negative impact on the environment. For this reason, it is important to establish a desired modal split that promotes the use of ecological means of transport. Previously, more environmentally friendly ways of travelling included walking, cycling and public transport (more in [1–5]). Now, changes also apply to the transport of cargo, in particular in city centers where cargo should be delivered chiefly by low emission vehicles. The process requires environmental responsibility of businesses and optimized supply chains [6–11]. The optimization process involves research and analysis of point and line components of the transport infrastructure and their traffic load (more in [12–15]). The improvements of the traffic flow distribution may shorten travelling time, reduce noise related to congestion, and indirectly mitigate the negative impact of transport (more in [16–21]).

Electromobility is a fairly new trend aimed at reducing the negative impact of transport on the environment. However, in many instances, the development of electromobility is too slow. From the economic, infrastructure and social points of view, electromobility is the result of the sustainable transport development policy, as specified in the White Paper and other policy documents describing various concepts in the EU (e.g. [22–24]). Key factors determining the development of electromobility is the potential and limitations of the transport system, as well as legal and economic issues pertaining to electromobility services. Specific features of a given region, including the

© Springer Nature Switzerland AG 2020
E. Macioszek and G. Sierpiński (Eds.): Modern Traffic Engineering in the System Approach to the Development of Traffic Networks, AISC 1083, pp. 99–113, 2020.
https://doi.org/10.1007/978-3-030-34069-8_9

potential and limitations of the transport system, are decisive regarding the possibility to develop electromobility services. Electromobility does not apply to means of transport only. It also covers the technical transport infrastructure and access to charging stations in a given area, as well as relevant technological solutions.

In the case of electric vehicles, a major limitation is the range the vehicle can cover on a fully charged battery. At the same time, the limited range is a factor that hampers the sales of electric vehicles. The advancement of the battery technology together with the growth in the number of charging stations are two major factors promoting the process. In cities of low electromobility level, particularly important is the adjustment of locations of the first-established charging stations. The process should initially focus on building trust to the use of electric vehicles. For this reason, locations of charging stations should correspond to the actual needs of users to promote electromobility and maximize its implementation effect [25].

A proper distribution of charging stations may seriously expedite the increase in the number of electric vehicles in cities. Research on the subject has been carried out, for instance, by the international project of "Electric travelling - platform to support the implementation of electromobility in Smart Cities based on ICT applications" [26] under the ERANET CoFund EMEurope programme financed among others by the National Centre for Research and Development.

The issue of selecting locations of charging stations can be examined using various methods. Such methods are listed in [27] depending on data availability and in [28] depending on the optimization algorithm used. The literature on the subject distinguishes various approaches based on data availability. Provided information on traffic flow is known, it is possible to determine the demand for charging based on traffic figures [29, 30]. The shortage of such information is a hindrance for the analysis of the transport network and land use [31, 32]. For instance, in [33] multiple domination models were used as an approach to the problem. It is based on a accessibility graph. Yet another method is based on an assumption that a vehicle needs to be stopped (parked) close to a public venue or home (facility that requires certain land use). This, however, necessitates relevant field studies followed by the use of variables for calibration in specific situations.

One of possibilities is to develop charging stations at existing petrol stations. This approach has been described in [34–36]. Depending on the motivation to travel, expectations of electric car users towards a charging station may vary. Charging on the route (petrol stations at motorways) requires superchargers. It is assumed, however, that in justified cases charging stations within urban areas may provide standard charging capacity (up to 23 kW). However, this necessitate the provision of a larger number of charging points.

The article analyzes a possible use of existing petrol stations as venues for charging electric vehicles. Such an approach is suggested for cities of low electromobility level. A case study covers the city of Sosnowiec and analyzes availability of potential locations coinciding with public venues.

2   Description of Sosnowiec

The study area presented in the article covers Sosnowiec. The city is located in the southern part of Poland, in the Silesia Province. It borders other cities, such as Katowice, Mysłowice, Będzin, Sławków, Jaworzno, Czeladź and Dąbrowa Górnicza (Fig. 1). The city consists of five districts that occupy 91 km^2 [37]. The city of Sosnowiec is the third largest in the province regarding its population after Katowice and Częstochowa. According to the National Statistical Office (GUS), at the end of 2018, the population of the city was 202 thousand [37].

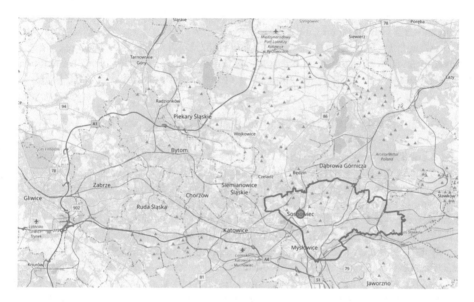

Fig. 1. Location of Sosnowiec in the Upper Silesia Conurbation (Source: [38])

Historically, Sosnowiec has been an industrial city. However, after the restructuring of industry and the liquidation of the majority of coal mines, the city shifted towards retail and services. More than 23 thou. businesses operate in the city [37], the majority of which are microenterprises representing the sector of services. Moreover, the city operates the Sosnowiec-Dąbrowa Subzone of the Katowice Special Economic Zone.

In Sosnowiec, the transport system is based on rail and road, long-distance coaches and public transportation. 25 km north of the city situated is the International Katowice Airport, whereas 70 km east from the city center operates the Kraków Airport.

The total length of the road system in the city is 333.5 km, including [39]:

- 12.0 km of national roads,
- 101.7 km of county roads, and
- 219.8 km of municipal roads.

In the city we can find expressways S1 and S86 (conveying some of the largest traffic in the country with over 100 thou. vehicles per day) and national roads DK1,

DK86, DK 79 and DK94. Moreover, the city road network of Sosnowiec includes over 26 petrol stations. Motorway A4 passes 9 km from the district of Śródmieście (City Center) to the Murckowska Junction and 1 km from the district of Niwka to the Mysłowice Junction. The favorable location on the junction of the main roads and railway lines promoted economic growth.

Sosnowiec is also one of the most polluted locations in Poland and Europe. The Silesian Agglomeration, which includes Sosnowiec, has the poorest air quality in the country (limit levels of PM 2.5, PM 10 and benzo(a)pyrene exceeded several hundred percent) [40].

One of the ways to reduce the negative impact of transport on the environment and improve the current situation is to introduce electromobility in the city. For instance, conventional individual cars should be replaced by electric ones. The Electromobility and Alternative Fuel Law adopted in Poland on 11th January 2018 requires cities to develop plans for establishing charging station until 2020 [41]. It is in line with the national guidelines for the development of electromobility adopted in 2016 [42].

3   Accessibility Analysis for Planned Charging Stations

The accessibility analysis for planned locations of charging stations has been based on assumptions specified in the document [36] which lists existing petrol stations as potential locations for charging infrastructure. The idea enables to use the traditional approach to the transport system, in particular among drivers, where petrol stations are considered as potential locations to charge batteries to the required level. The use of existing parking space and fitting charging points at those petrol stations can significantly reduce the cost of the process. Considering plans to change the percentage of electric vehicles on our roads in the nearest future, it should enable to replace petrol distributors with charging equipment, including fast chargers. Moreover, the traditional approach to the use of petrol stations by petrol and Diesel driven vehicles is going to be expanded with alternative fuels, including electricity.

The article further presents the analysis of distance between public venues of specified functions and existing petrol stations. The analysis uses QGIS 2.12 Lyon and Hub Distance functionality of the MMQGIS plug, which enables to determine the nearest neighbors in two predefined layers of points. Data on petrol stations and public facilities have originated from the Integrated System of Spatial Information, a database of the City of Sosnowiec [43]. The Center of Gravity and Geoprocessing Tools have been used to determine point layers that include locations of facilities to be further examined.

The analysis also distinguishes between various functions of public facilities as specified in the Integrated System of Spatial Information:

- retail and service,
- education,
- office,
- medical care, and
- industry.

Below presented are findings of the analysis for the above mentioned types of facilities.

3.1 Access to Planned Charging Points from Retail and Service Facilities

The first and the largest category distinguished includes retail and service facilities. At the same time, it should be noted that customers stay in those facilities for a short time only. The graphic interpretation of the distance function for retail and service facilities and existing petrol stations is shown in Fig. 2.

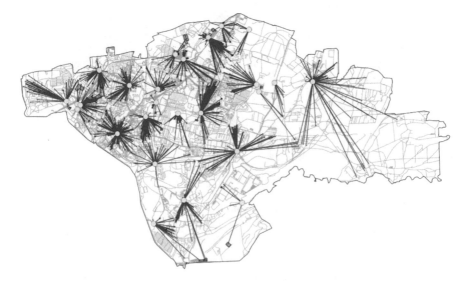

Fig. 2. Distance between charging points and retail and service facilities

The number of retail and service facilities in Sosnowiec covered by the analysis is 1378. More than 78% of those facilities are located less than 1 km from the nearest petrol station. In less than 4% of cases the distance exceeds 1.5 km. A detailed distribution of retail and service facilities and nearest petrol stations is shown in Fig. 3.

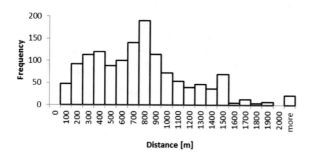

Fig. 3. Distance between retail and service facilities and nearest petrol stations

3.2   Access to Planned Charging Points from Education Facilities

Another group of facilities vary in terms of handling time. The educational function includes establishments for children and facilities providing training courses for adults. The graphic interpretation of the distance between educational, science and culture and sport facilities and existing petrol stations is shown in Fig. 4.

Fig. 4. Distance between charging points and education facilities

The analysis covered 234 public buildings with educational function in the city of Sosnowiec. Less than 80% of such facilities are located within 1 km from the nearest petrol station. A detailed distribution of facilities with educational function, including science and culture and sports, and nearest petrol stations has been shown in Fig. 5.

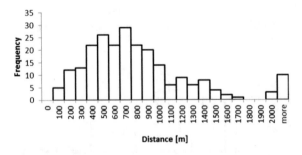

Fig. 5. Distance between educational facilities and nearest petrol stations

3.3 Access to Planned Charging Points from Office Facilities

Office buildings, which is yet another group of facilities, attract customers to a lesser degree. In the case of office buildings, we may assume that only employees will come to stay longer. The graphic interpretation of the distance between office facilities and existing petrol stations is shown in Fig. 6.

Fig. 6. Distance between charging points and office facilities

Sosnowiec has 352 office facilities, of which 79% are located less than 1 km from the nearest petrol station. A detailed distribution of office facilities and nearest petrol stations has been shown in Fig. 7.

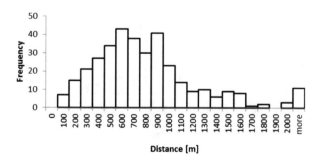

Fig. 7. Distance between office facilities and nearest petrol stations

3.4   Access to Planned Charging Points from Medical Care Facilities

The graphic interpretation of the distance between medical care facilities, such as outpatient clinics or hospitals and existing petrol stations is shown in Fig. 8.

Fig. 8. Distance between charging points and medical care facilities

The analysis covered 88 healthcare facilities, including hospitals and outpatient clinics, in the city of Sosnowiec. 78% of those facilities are situated less than 1 km from nearest petrol stations. A detailed distribution of the distance between medical care facilities and petrol stations is shown in Fig. 9.

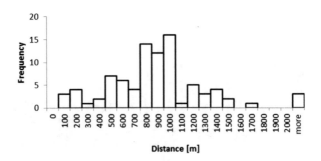

Fig. 9. Distance between healthcare facilities and nearest petrol stations

3.5   Access to Planned Charging Points from Industrial Facilities

The final group examined includes industrial facilities. The transport service to those facilities requires direct access (in particular to warehouses). Usually, people who stay longer at those facilities include employees. The graphic interpretation of the distance between industrial facilities, including warehouses, tanks and silos, and existing petrol stations has been shown in Fig. 10.

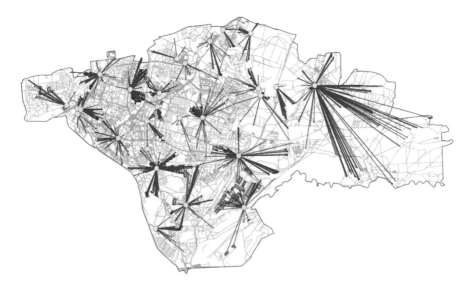

Fig. 10. Distance between charging points and industrial facilities

The analysis covered 926 industrial facilities in the city of Sosnowiec. 66% of them are located less than 1 km from the nearest petrol station. It is the lowest result in the overall accessibility analysis. At the same time, it should be noted that industrial facilities are often situated at the outskirts of residential areas. Thus, an industrial area has poorer transport infrastructure. A detailed distribution of distances between industrial facilities, including production halls, warehouses, tanks and silos, and nearest petrol stations has been shown in Fig. 11.

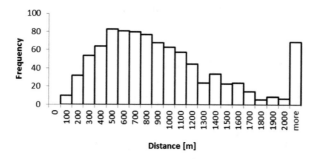

Fig. 11. Distance between industrial facilities and nearest petrol stations

3.6   Accessibility of Selected Petrol Stations

The analysis of the access to electromobility services for the city of Sosnowiec has been carried out by identifying two petrol stations depending on their location in the core city center (I) and suburban housing area (II). The graphic interpretation of the distance between facilities of various functions and their nearest petrol stations is shown in Fig. 12. Moreover, the same figure shows two petrol stations to be further examined.

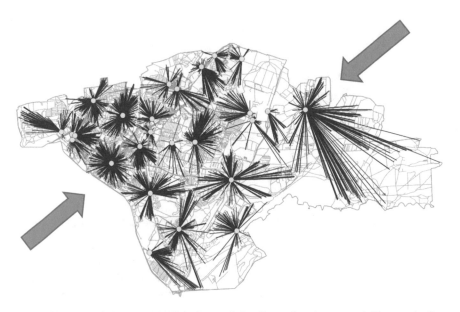

Fig. 12. Accessibility of potential places of charging points (nearest neighbor method)

In the core city center, Sosnowiec has 237 facilities of various function in the nearest neighborhood of petrol station I, of which 67% are facilities located within 600 m and 31% up to 1 km. Details are shown in Fig. 13.

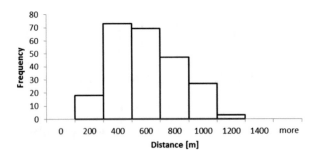

Fig. 13. Accessibility of petrol station I in the core city center

In the case of petrol station II, there are 237 facilities of various functions in the vicinity, of which 32% are facilities located up to 1 km away, 25% of facilities within 2 km, and 19% up to 3 km. Details are shown in Fig. 14.

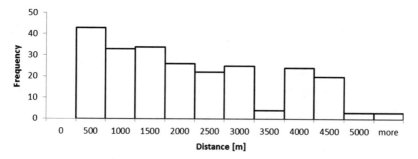

Fig. 14. Accessibility of petrol station II in suburban housing area

There is no strict relationship between the location of petrol station II and facilities in its vicinity. At the same time, we can see (Fig. 12) a shortage of petrol/charging stations in the eastern part of the city.

3.7   Accessibility of Petrol Stations - Statistical Analysis

Finally, the analysis focused on figures for facilities of various types. Table 1 presents statistics regarding distances between petrol stations and facilities in Sosnowiec with breakdown by their functions.

Table 1. Statistics on accessibility and distances [m] between petrol stations and facilities with breakdown by their functions.

Measure	Retail and service	Education	Office	Medical care	Industry, warehouses	All
Total number	1378	234	352	88	926	2978
Average	726.87	780.79	791.96	857.88	979.08	821.1
Standard deviation	482.41	609.9	619.06	529.18	818.34	640.58
Kurtosis	8.51	13.24	20.37	14.99	7.06	12.09
Minimum	1.78	50.82	5.22	11.26	37.78	1.78
Maximum	4208.69	4350.64	5941.48	4044.32	5004.54	5941.48
Quantile 1	375.18	425.47	460.89	556.61	481.71	433.38
Quantile 2	695.67	664.5	683.1	831.66	778.45	709.76
Quantile 3	928.18	938.6	936.53	958.32	1150.47	1007.04
Variability	0.66	0.78	0.78	0.62	0.84	0.78
Skewness	1.89	2.96	3.58	2.81	2.45	2.82

The average distance to a petrol station within 700–800 m is considered favorable in terms of potential establishment of charging stations for electric vehicles.

The analysis of the asymmetric distribution of data for specific groups of facilities helped to determine the skewness factor. All factors have a positive value, which means that the majority of data are below the average value. It is compensated by a few elements with high positive values [44]. At the same time, the major asymmetry has been found in the group of office facilities. The degree of concentration has been determined using kurtosis. Values determined show major concentration in the case of office facilities. To highlight data dispersion, the analysis determined a standard deviation and variability factor presented as a deviation per average unit [44] as follows (1):

$$W = S/\bar{x} = \sqrt{\frac{1}{n-1} \cdot \sum_{i=1}^{n} (x_i - \bar{x})^2} \left/ \frac{1}{n} \cdot \sum_{i=1}^{n} x_i \right. \tag{1}$$

The value of the variability factor (Table 1) shows heterogeneity of the set examined.

4   Conclusions and Further Research

The method may support decision-making while determining the location for the first charging stations to be established in a given area. The habit among drivers to use existing petrol stations should help to reduce anxiety related to use of a new type of vehicles. At the same time, depending on the distribution of petrol stations in the city road network, one may decide to install additional charging stations in areas of poorer accessibility to petrol stations (too large walking distance from petrol station to destination).

The preliminary studies have been limited to the analysis of accessibility of petrol stations as potential locations for charging installations. The decision to maintain or liquidate existing petrol stations is a separate issue. At an early stage of developing electromobility in a specific area, it is advisable to maintain both solutions available. One the one hand, it may slow down the development but, on the other hand, it will mitigate resistance among users of traditional diesel/petrol driven vehicles. Yet another issue that may be further examined is the supply of power needed at charging stations. Although the primary source of electricity should be the grid [45, 46], the literature describes other solutions, such as mobile batteries (similar to power banks used for mobile devices) or photovoltaic systems (e.g. [47–49]).

Acknowledgements. The present research has been financed from the means of the National Centre for Research and Development as a part of the international project within the scope of ERA-NET CoFund Electric Mobility Europe Programme "Electric travelling - platform to support the implementation of electromobility in Smart Cities based on ICT applications".

References

1. Jacyna, M., Żak, J., Jacyna-Gołda, I., Merkisz, J., Merkisz-Guranowska, A., Pielucha, J.: Selected aspects of the model of proecological transport system. J. KONES Powertrain Transp. **20**, 193–202 (2013)
2. Lejda, K., Mądziel, M., Siedlecka, S., Zielińska, E.: The future of public transport in light of solutions for sustainable transport development. Sci. J. Sil. Univ. Technol. Ser. Transp. **95**, 97–108 (2017)
3. Turoń, K., Czech, P., Juzek, M.: The concept of walkable city as an alternative form of urban mobility. Sci. J. Sil. Univ. Technol. Ser. Transp. **95**, 223–230 (2017)
4. Celiński, I.: Using GT planner to improve the functioning of public transport. In: Macioszek, E., Sierpiński, G. (eds.) Recent Advances in Traffic Engineering for Transport Networks and Systems. LNNS, vol. 21, pp. 151–160 (2018)
5. Galińska, B.: Intelligent decision making in transport. evaluation of transportation modes (types of vehicles) based on multiple criteria methodology. In: Sierpiński, G. (ed.) Integration as Solution for Advanced Smart Urban Transport Systems. AISC, vol. 844, pp. 161–172 (2019)
6. Turoń, K., Golba, D., Czech, P.: The analysis of progress CSR good practices areas in logistic companies based on reports "Responsible Business in Poland. Good Practices" in 2010–2014. Sci. J. Sil. Univ. Technol. Ser. Transp. **89**, 163–171 (2015)
7. Golba, D., Turoń, K., Czech, P.: Diversity as an opportunity and challenge of modern organizations in TSL area. Sci. J. Sil. Univ. Technol. Ser. Transp. **90**, 63–69 (2016)
8. Kauf, S.: City logistics - a strategic element of sustainable urban development. Transp. Res. Procedia **16**, 158–164 (2016)
9. Ocicka, B., Wieteska, G.: Sharing economy in logistics and supply chain management. LogForum **13**(2), 183–193 (2017)
10. Wątróbski, J., Małecki, K., Kijewska, K., Iwan, S., Karczmarczyk, A., Thompson, R.G.: Multi-criteria analysis of electric vans for city logistics. Sustainability **9**(8), 1–34 (2017)
11. Żak, J., Galińska, B.: Design and evaluation of global freight transportation solutions (corridors). Analysis of a real world case study. Transp. Res. Procedia **30**, 350–362 (2018)
12. Małecki, K.: The importance of automatic traffic lights time algorithms to reduce the negative impact of transport on the urban environment. Transp. Res. Procedia **16**, 329–342 (2016)
13. Macioszek, E., Lach, D.: Analysis of the results of general traffic measurements in the west pomeranian voivodeship from 2005 to 2015. Sci. J. Sil. Univ. Technol. Ser. Transp. **97**, 93–104 (2017)
14. Santos, G.: Road transport and CO2 emissions: what are the challenges? Transp. Policy **59**, 71–74 (2017)
15. Macioszek, E., Lach, D.: Comparative analysis of the results of general traffic measurements for the Silesian Voivodeship and Poland. Sci. J. Sil. Univ. Technol. Ser. Transp. **100**, 105–113 (2018)

16. Cárdenas, O., Valencia, A., Montt, C.: Congestion minimization through sustainable traffic management. A micro-simulation approach. LogForum **14**(1), 21–31 (2018)

17. Celiński, I.: Evaluation method of impact between road traffic in a traffic control area and in its surroundings. Ph.D. Dissertation. Oficyna Wydawnicza Politechniki Warszawskiej, Warsaw (2018). (in polish)

18. Celiński, I.: GT planner used as a tool for sustainable development of transport infrastructure. In: Suchanek, M. (ed.) New Research Trends in Transport Sustainability and Innovation. SPBE, pp. 15–27. Springer, Cham (2018)

19. Galińska, B.: Multiple criteria evaluation of global transportation systems - analysis of case study. In: Sierpiński, G. (ed.) Advanced Solutions of Transport Systems for Growing Mobility. AISC, vol. 631, pp. 155–171. Springer, Cham (2018)

20. Macioszek, E., Lach, D.: Analysis of traffic conditions at the Brzezinska and Nowochrzanowska intersection in Myslowice (Silesian Province, Poland). Sci. J. Sil. Univ. Technol. Ser. Transp. **98**, 81–88 (2018)

21. Pijoan, A., Kamara-Esteban, O., Alonso-Vicario, A., Borges, C.E.: Transport choice modeling for the evaluation of new transport policies. Sustainability **10**, 1–22 (2018)

22. European Commission: White Paper: Roadmap to a Single European Transport Area - Towards a competitive and resource efficient transport system. https://ec.europa.eu/transport/sites/transport/files/themes/strategies/doc/2011_white_paper/white-paper-illustrated-brochure_en.pdf

23. European Commission: White Paper on the Future of Europe. Reflections and scenarios for the EU27 by 2025. https://ec.europa.eu/commission/sites/beta-political/files/white_paper_on_the_future_of_europe_en.pdf

24. European Commission: Communication From the Commission to the European Parliament, the Council, the European Economic and Social Committee and the Committee of the Regions: Clean Power for Transport: a European alternative fuels strategy. https://eur-lex.europa.eu/LexUriServ/LexUriServ.do?uri=COM:2013:0017:FIN:EN:PDF

25. Sierpiński, G.: Support for areas with low level of electromobility through the use of big data. In: Proceedings of the 5th World Congress on New Technologies (NewTech 2019). ICERT, NewTech, Lisbon, pp. 106-1–106-4 (2019)

26. Electric travelling - Platform to support the implementation of electromobility in Smart Cities based on ICT applications - Project proposal under EMEurope programme (2016)

27. Csiszár, C., Csonka, B., Földes, D., Wirth, E., Lovas, T.: Urban public charging station locating method for electric vehicles based on land use approach. J. Transp. Geogr. **74**, 173–180 (2019)

28. Shareef, H., Islam, M.M., Mohamed, A.: A review of the stage-of-the-art charging technologies, placement methodologies, and impacts of electric vehicles. Renew. Sustain. Energy Rev. **64**, 403–420 (2016)

29. Alhazmi, Y.A., Mostafa, H.A., Salama, M.M.A.: Optimal allocation for electric vehicle charging stations using trip success ratio. Int. J. Electr. Power Energy Syst. **91**, 101–116 (2017)

30. Qiao, Y., Huang, K., Jeub, J., Qian, J., Song, Y.: Deploying electric vehicle charging stations considering time cost and existing infrastructure. Energies **11**, 1–13 (2018)

31. Gkatzoflias, D., Drossinos, Y., Zubaryeva, A., Zambelli, P., Dilara, P., Thiel, Ch.: Optimal allocation of electric vehicle charging infrastructure in cities and regions. Joint Research Centre. Science for Policy Report. European Commission, Ispra (2016)

32. Erbaş, M., Kabak, M., Özceylan, E., Çetinkaya, C.: Optimal siting of electric vehicle charging stations: a GIS-based fuzzy multi-criteria decision analysis. Energy **163**, 1017–1031 (2018)

33. Gagarin, A., Corcoran, P.: Multiple domination models for placement of electric vehicle charging stations in road networks. Comput. Oper. Res. **96**, 69–79 (2018)

34. Saarijärvi, E.: SGEM 3.3.6-1: Gas station conversion to EV fast charging station. Aalto University School of Science and Technology, Department of Electrical Engineering, Espoo (2010)
35. Bi, R., Xiao, J., Viswanathan, V., Knoll, A.: Influence of charging behaviour given charging station placement at existing petrol stations and residential car park locations in Singapore. Procedia Comput. Sci. **80**, 335–344 (2016)
36. Fernández, G., Torres, J., Cervero, D., García, E., Alonso, M.A., Almajano, J., Machín, S., Bludszuweit, H.: EV charging infrastructure in a petrol station, lessons learned. In: Proceedings of the 2018 International Symposium on Industrial Electronics - INDEL, Banja Luka, pp. 1–6. IEEE Press (2018)
37. General Statistical Office. http://stat.gov.pl
38. Open Street Map. https://www.openstreetmap.org/
39. The local revitalization program of the city of Sosnowiec for 2010–2020. http://www.sosnowiec.pl/_upload/file/Lokalny%20Program%20Rewitalizacji%20miasta%20Sosnowca%20na%20lata%202010-2020.pdf
40. D4.16 Mutual Learning Workshop Analysis Report. ClairCity: Citizen-led air pollution reduction in cities. Grant agreement No. 689289. http://www.claircity.eu/wp-content/uploads/2019/01/D4.14-Mutual-Learning-Workshop-complete-First-City.pdf
41. Electromobility and Alternative Fuel Act of 11th January 2018. http://prawo.sejm.gov.pl/isap.nsf/download.xsp/WDU20180000317/T/D20180317L.pdf
42. Ministry of Energy: Development Plan for Electromobility in Poland. Energy for the future. Ministry of Energy. https://www.gov.pl/documents/33372/436746/DIT_PRE_PL.pdf/ebdf4105-ef77-91df-0ace-8fbb2dd18140
43. Integrated Spatial Information System of the City of Sosnowiec. http://www.zsip.sosnowiec.pl
44. Bruce, P., Bruce, A.: Practical Statistics for Data Scientists. O'Reilly Media Incorporation, Sebastopol (2017)
45. Richardson, D.B.: Electric vehicles and the electric grid: A review of modeling approaches, impacts, and renewable energy integration. Renew. Sustain. Energy Rev. **19**, 247–254 (2013)
46. Yong, J.Y., Ramachandaramurthy, V.K., Tan, K.M., Mithulananthan, N.: A review on the state-of-the-art technologies of electric vehicle, its impacts and prospects. Renew. Sustain. Energy Rev. **49**, 365–385 (2015)
47. Nunes, P., Figueiredo, R., Brito, M.C.: The use of parking lots to solar-charge electric vehicles. Renew. Sustain. Energy Rev. **66**, 679–693 (2016)
48. Alghoula, M.A., Hammadib, F.Y., Aminc, N., Asimb, N.: The role of existing infrastructure of fuel stations in deploying solar charging systems, electric vehicles and solar energy: a preliminary analysis. Technol. Forecast. Soc. Chang. **137**, 317–326 (2018)
49. Volkswagen Group News: Electrifying World Premiere: Volkswagen offers First Glimpse of Mobile Charging Station. https://www.volkswagen-newsroom.com/en/press-releases/electrifying-world-premiere-volkswagen-offers-first-glimpse-of-mobile-charging-station-4544

Study of Characteristics of Road Traffic Streams in Pedestrian Crossing - Affected Areas

Ireneusz Celiński[(✉)]

Faculty of Transport, Silesian University of Technology, Katowice, Poland
ireneusz.celinski@polsl.pl

Abstract. The article addresses the problem of studying the characteristics of traffic (including pedestrian) streams in areas affected by pedestrian crossings. Relations between these characteristics determine interactions between vehicular and pedestrian streams, the consequences of which include variable parameters of vehicle queues emerging in pedestrian crossing-affected areas, delay, difficulties crossing the road by pedestrians, etc. Such consequences may also include road incidents and accidents (knocking down pedestrians, rear-end collisions with the preceding vehicle, etc.). The problem in question is important for various reasons, including the share of road accidents involving pedestrians in the Polish road network remaining invariably high, and even despite some decisive actions aimed to counteract such incidents.

Keywords: Pedestrian crossing · Characteristics of traffic streams · Traffic detectors

1 Introduction

Road accidents involving pedestrians still account for a significant percentage of the total number of such incidents occurring in the Polish road network. According to official statistics (KGP, Police Headquarter in Warsaw, 2016), they exceed 30% of the total number of accidents registered to have happened in the Polish road network [1]. Most such accidents take place in the direct vicinity of pedestrian crossings as well as in their surroundings. The area of the surroundings is determined by the range of impact of pedestrian crossings. It can be specified by the length of vehicle queues or by measuring the variability of such characteristics as those presented in this article. From the perspective of the accident statistics observed, it is important to investigate the reasons for such incidents happening in the road network in pedestrian crossing-affected areas. The reasons for the road accidents involving pedestrians may vary, including inattention of pedestrians, technical defects of vehicles, bad weather conditions, failure to adapt the vehicle speed to traffic conditions, etc. These reasons may also be correlated with the characteristics discussed in this article as well as the relationships between them.

© Springer Nature Switzerland AG 2020
E. Macioszek and G. Sierpiński (Eds.): Modern Traffic Engineering in the System Approach to the Development of Traffic Networks, AISC 1083, pp. 114–130, 2020.
https://doi.org/10.1007/978-3-030-34069-8_10

In the context of the article's subject matter and the traffic studies addressed below, it is possible to analyse the potential impact of a system of characteristics of the pedestrian and vehicular traffic streams on the possibility of road accidents happening.

In order to investigate the characteristics of road and pedestrian traffic, the studies were conducted in a small town (Mikołów, Poland), in an area affected by pedestrian crossings. A specific street where many events of this type occurred over the last decade (2009÷2019), including several fatal accidents, was analysed.

The research on basic road traffic characteristics (traffic volume, density, length of vehicle queues, delay, etc.) was extensively addressed in the literature of the subject of the 1940s, 50s and 60s. In the 1960s, traffic research started to be conducted on a large scale as some significant traffic issues had surfaced. Such research is also consistently performed at this day and age [2–7]. Why is it reasonable to repeatedly conduct traffic studies? It is so because drivers' behaviour patterns change significantly over the years, country, region, age thus affecting the characteristics of the vehicular traffic stream. New types of vehicles with increasingly advanced power transmission systems are introduced. In certain metropolitan areas and agglomerations, driving styles of vehicle drivers are notably different than those observed in other urban and rural areas (capital city Warsaw vs country). Certain driving style patterns are imitated across the country but gradually (very slow). What also changes is the road network infrastructure. Roundabouts are becoming increasingly popular in the Polish road network (including a turbine roundabout), and in some towns in Poland they already account for a significant percentage of road intersections (e.g. Tychy, Mikołów, Rybnik, Ruda Śląska, etc.). And most importantly, also road traffic regulations and traffic organisation change (relatively often). These and other changes constitute the rationale for systematic studies of road traffic characteristics. They are assumed to confirm the current state of knowledge concerning these problems, and to update it in the scope of the observed changes to the road traffic characteristics in question. The technological development of microprocessor-based devices and the fact that their costs are declining make it possible to conduct such research on a nearly global basis.

An example of such a cheap device is the one used in the studies addressed in this article. With a very low production cost (ca. PLN 50÷100 in 2018, equal ca. 12,5÷25,0 €), this device can be installed in a quasi-continuous manner in the road network space. By that means, it is possible to control traffic characteristics in the road network on a global scale, and consequently also to control delay in the road network more effectively. Presented moving object detector differentiates the presence of vehicle from presence of pedestrian at the pedestrian crossing by preliminary measurement made using model detector with counter. The model detector has an moving object counter and is used to configure the location of ordinary detectors so that the detection field covers its area only in an appropriate range. This requires conducting preliminary studies of the best location. The principle is that one detector observes one stream of traffic (pedestrian or road).

What was analysed in the study besides the road traffic in the vicinity of passenger crossings was also the pedestrian traffic characteristics. In the future, such a study may allow a correlation to be found between traffic characteristics at pedestrian crossings and their impact on the probability of road accidents and incidents. This article is mainly limited to traffic studies and a discussion of the research tool used to conduct the relevant analyses. Another problem related to the subject matter is that of vehicle queues forming at pedestrian crossings, and more specifically their relation to the characteristics of vehicular and pedestrian traffic streams. The manner in which traffic is organised at pedestrian crossings (with and without traffic lights, underground crossings) should generally take the characteristics of vehicular and pedestrian traffic into consideration [2]. The methods typically applied to date when designing pedestrian crossings mainly consider two parameters, i.e. the speed of vehicular traffic and the traffic stream volume. Even so, the use of a specific pedestrian crossing type is determined by certain fluctuations in the traffic stream characteristics, which can be studied using the technique discussed in this article.

2   Traffic Stream Studying Technique

The characteristics of vehicular and pedestrian traffic streams were studied using a device intended for traffic detection of the author's own design. It detects the presence of a vehicle or a pedestrian (or any moving object) in the road network cross-section, making it possible to record the event detection time (one devices assigned to one stream). The device is based on a microwave motion detector (RCWL-0516) coupled with a GPS module (NEO6) and an SD memory card. It is comprehensively managed by means of the ATmega class microcontroller. A photograph of a model device, a schematic diagram of links between individual modules and a sample field testing station have been shown in Fig. 1 a–c.

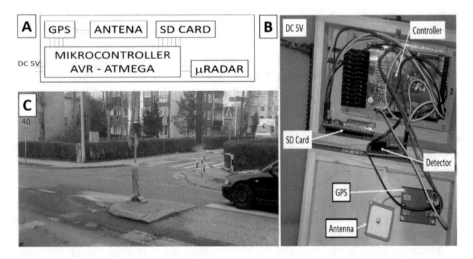

Fig. 1. Vehicle detector and measuring station: (a) schematic diagram of the device, (b) photograph showing the device installed, (c) sample testing station in the road network

The device shown in Fig. 1a and b enables detection of a car or a pedestrian at a distance of ca. 5 m away from the device (distance from microwave detector's front surface). When appropriately positioned (via model detector) in a street cross-section, the device records the presence of a vehicle or a pedestrian (binary values). The vehicle presence is recorded with the time accuracy conforming to the GPS standard by using the latter system's receiver connected with the microcontroller. The detection data are stored on an SD memory card, making it possible to record at least 24 h of continuous traffic stream monitoring ($\Delta t = 1$ s).

The study consisted in observing pedestrian crossings in a street characterised by a high rate of accidents involving pedestrians, including fatalities (several such cases over the last 10 years). Following a series of fatal accidents, some of them were equipped with pedestrian refuges (also referred to as refuge islands). A schematic representation of the street where the study was conducted has been provided in Fig. 2a.

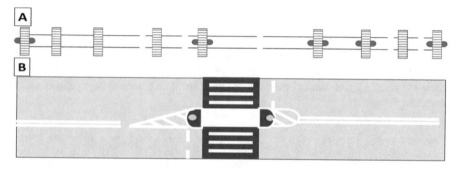

Fig. 2. Overview of the street showing the pedestrian crossings: (a) space layout of pedestrian crossings, (b) schematic diagram of a single pedestrian crossing with a refuge island

The results obtained in the study were used to describe the characteristics of vehicular and pedestrian traffic streams (number of detection entries within a pre-defined time interval, headway, distribution of vehicle detection instants in time), which in turn were used to simulate the process in which these traffic structures exert their impact. The simulations were performed using a proprietary program which enables simulation of traffic streams, events and accidents involving pedestrians. The traffic streams were studied at rush hours (7.00–8.00, 14.00–15.00).

The study focused on rare streams (maximum disorder process) characterised by high randomness being observed, a small number of entries in short intervals, and the time interval between entries being exponential (for pedestrians) or shifted exponential in road traffic. What can be observed in the vehicular traffic is a safety interval, which was not measured in the study in question.

3 Selected Results of Traffic Stream Measurements

Figure 3 illustrates sample data concerning the detected presence of a vehicle in the street cross-section obtained in the study of the characteristics of vehicular traffic streams. The vehicle detection characteristics have been presented for two directions of vehicular traffic (described in the geocentric system: east and west). The vehicle detection results shown were synchronised with the GPS system time (via the GPS receiver), and the consecutive values on the X-axis correspond to consecutive seconds of measurement (one by one). Which means, that: every presence of the vehicle measurement can occur with a one second interval when the time of the signal registered in the GPS receiver changes (it measures by atomic clock of the satellite). On the Y-axis, the value 1 means that the vehicle was detected in the measuring cross-section. The value zero means that the vehicle wasn't detected in the measuring cross-section. The device does not specify intermediate values from range (0.1). However, such a measurement can be very interesting. It should be noted that in some cases several adjacent peaks (ones) in Fig. 3a and b indicate the instant of detection of the same vehicle, depending on the current parameters of the stream (mainly speed and type structure). Such characteristics can be processed automatically or manually. This article comments upon manual calculations performed in a spreadsheet (this is a time-consuming approach).

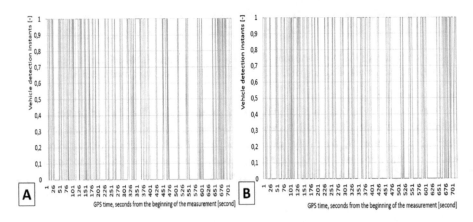

Fig. 3. Characteristics of vehicular traffic; vehicle detection instants direction: (a) eastward, (b) westward

As shown in Fig. 3, the distribution of vehicle detection instants (presence) was uneven (random). This is the case of both traffic directions, and there were no significant trends observed in terms of the uniformity of this distribution.

Figure 4 shows superimposed time instants when vehicles running in both traffic directions (east and west) were detected at a pedestrian crossing. Between these time points, there are *time windows* where the crossing occupancy time can be used by the pedestrians. Figure 5 shows the time intervals between successive vehicles, referred as to headway, for the two traffic directions analysed. The headway has been described as an interval distribution by disregarding the safety interval between vehicles (this characteristic was not measured, hence the interval of 0–2). On the X-axis, class intervals corresponding to the respective headway ranges have been plotted, while on the Y-axis - the numbers recorded in these intervals.

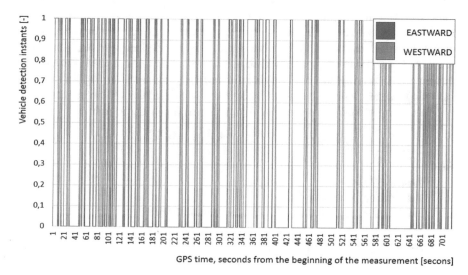

Fig. 4. Superimposed vehicle detection times for both traffic directions

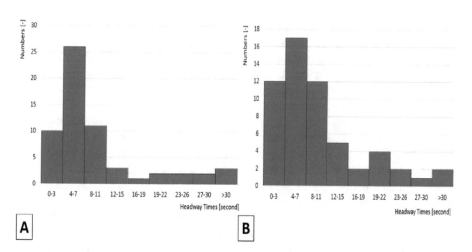

Fig. 5. Characteristics of traffic; headway, direction: (a) eastward, (b) westward

Such traffic characteristics as those shown in Fig. 4 can mainly be interpreted in statistical terms (e.g. average number of entries), since in many cases it is difficult to achieve visually. Unlike in Fig. 4, the characteristics presented in Fig. 5 show significant differences between traffic streams moving in both directions. What differs is not only the number of vehicles running in both directions, but also the observed headway values. Differences between characteristics are mainly observed in classes above the headway time of 11 s (i.e. outside the typical critical gap time values observed in real-life conditions). In both cases, the predominant headway was between 4 and 7 s. The most common time values ranged between 2 and 11 s. It should be noted that there is no buffer (safety interval between vehicles, referred to as distribution memory) marked on the graph, because in fact it is a shifted exponential distribution, where the lowest observed headway time values were 2 s. (the measuring device is not suitable for measuring the buffer value for accuracy reasons). In Fig. 5, the range containing values close to the buffer <0.3> have been intentionally highlighted.

The density of shifted exponential distribution is given by the formula:

$$f(x) = \lambda e^{-\lambda(x-\Delta)}, \quad x \geq \Delta, \lambda > 0, \Delta > 0 \tag{1}$$

where:

x - headway [s],
λ - distribution parameter [-],
Δ - safety interval (or buffer, distribution memory) [s].

When correlated with the λ parameter of the headway distribution (mean), the safety interval can be a measure of even spacing in traffic stream. If the safety interval value is nearing the average distribution value, such a stream is *weakly random*. Instead, it is rather evenly spaced. Figure 5 a and b imply that traffic stream studies are random in nature with regard to the headway. For the "eastward" stream, the arithmetic mean is 9.88 s, while for the "westward" stream −9.64 s. In both cases, the mean exceeds the buffer value of less than 2 s by nearly 5 times. Therefore, one can observe a high degree of randomness in vehicle detection entries in both traffic streams.

The results observed within the range of headway time intervals between vehicles are consistent with the relevant literature data concerning traffic stream parameters at non-signal-controlled intersections [2–4, 11, 12].

Figure 6 shows sample characteristics of pedestrian presence detection at one of the pedestrian crossings. Pedestrian traffic was detected simultaneously in both directions (hence the possible gap of 0 s between them). The study did not reveal any cases of pedestrians passing each other at the microwave detector's front surface. The figure shows moderate irregularity of the pedestrian traffic. The X-axis corresponds to consecutive seconds during the traffic measurement.

Figure 7 presents the headway between the pedestrians participating in road traffic at the pedestrian crossing studied. The predominant value of pedestrian headway ranges between 4 and 7 s. It should be noted that, at this particular crossing, this range is identical with that of the vehicular traffic. Synchronisation (its lack) of the vehicular and pedestrian traffic streams in time may cause queues. Figure 7 also shows other values of the interval between pedestrians which depart from the mean pedestrian headway a rather extreme manner, as the differences come to several dozen seconds. Outside rush hours, one should expect more of such extreme observations. These cases,

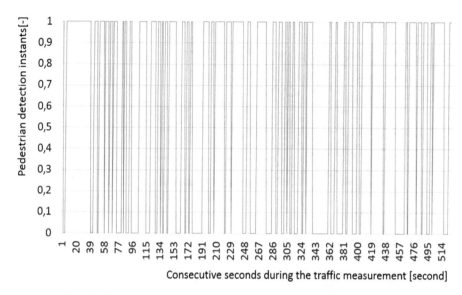

Fig. 6. Characteristics of pedestrian traffic, pedestrian detection instants

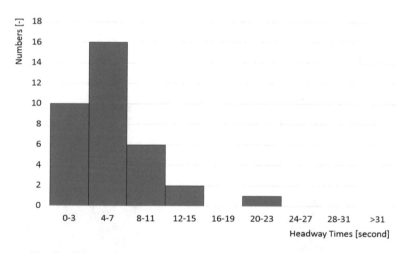

Fig. 7. Characteristics of pedestrian traffic, headway between pedestrians

however, constitute a problem of slightly different nature compared to those discussed in this article (e.g. these situations may correspond to a sudden pedestrian incursion onto the roadway).

Figure 8 shows the distribution of the number of pedestrians recorded within a 20-s interval (the same interval would later be assumed to analyse the distribution of the number of entries in the vehicular traffic analysis). In this case, Fig. 8 indicates the number of pedestrians, and not the headway, on the X-axis.

The data acquired in the aforementioned manner, concerning the distribution of the number of entries for both pedestrian and vehicular traffic, were verified in terms of

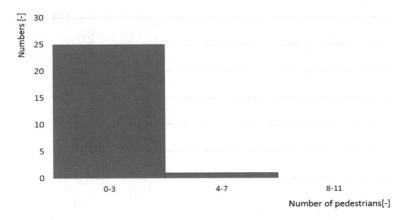

Fig. 8. Characteristics of pedestrian traffic, number of pedestrians in 20-s intervals

their compliance with the Poisson distribution (one can also study binominal or other discrete distribution), assuming the following probability distribution:

$$P(X = k) = \frac{e^{-\lambda}\lambda^k}{k!}, \quad k = 0, 1, 2, 3, \ldots \tag{2}$$

where:

k - number of detection entries [-],

λ - distribution parameter [-]

What the author had assumed for the sake of the study (as for the pedestrian traffic stream) was the analysis interval of 20 s. Figure 9 shows the distribution of the number of vehicle detection entries within 20-s intervals.

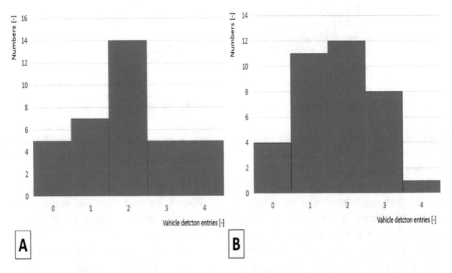

Fig. 9. Characteristics of vehicular traffic; vehicle detection entries: (a) eastward, (b) westward

The pre-set measurement interval (20 s) is dominated by the values of two vehicle detection entries. Most of the observations range between zero entries and 4 entries. The $\chi2$ test was conducted at the significance level of $\alpha = 0.05$, and the hypothesis on the Poisson distribution of the number of detection entries in both cases of vehicular traffic stream was positively verified (Table 1). It should be noted that, while calculating the number of detection entries, one could also find vehicles exactly at the boundary of the 20-s range (which could not be unambiguously included in a single range; ca. 10% of total measurements on average). What the author also disregarded was the ends of the measurement period exceeding the last interval. The problem of measuring accuracy

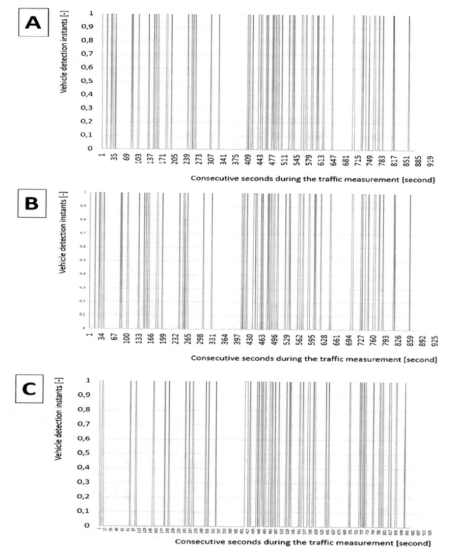

Fig. 10. Characteristics of vehicular traffic for three cross-sections in one street: a–c/consecutive measurement cross-section 1, 2, 3

($\Delta = 1$ s) needs to be solved in the course of further research - perhaps by changing the resolution of the time base (RTC system clock instead of GPS).

The characteristics discussed above (Figs. 3, 4, 5, 6, 7, 8 and 9) pertain to the selected single cross-section of the road network or a pedestrian crossing (although for two traffic directions). It is interesting how they change in consecutive road network cross-sections, e.g. for a single street. According to the research results elaborated in [8, 9] for the railway network, these characteristics are constant for a single route. In order to investigate how the traffic streams characteristics change in a series of streets subject to examination, the relevant measurements were compared using 3 consecutively arranged detectors. The sensors were installed in three different street cross-sections at a distance of ca. 15 m from one another (mean length of three Passenger Car Equivalent/Unit type vehicles including ca. 0.5 m buffers at the vehicle's back and front) The vehicle detection time recorded by the individual detectors have been shown in Fig. 10.

Figure 10 has revealed a far greater irregularity in the number of vehicle detection entries compared to a single cross-section of the same street (breaking procedure before pedestrian). What has mainly changed is the number of vehicles (exiting to private access roads past cross-section no. 1) and headway, and the latter change is considerable. In terms of the characteristics of vehicle detection in time, cross-section no. 1 also deviates significantly from cross-sections no. 2 and 3. Figure 11 presents the headway between vehicles observed in these three cross-sections.

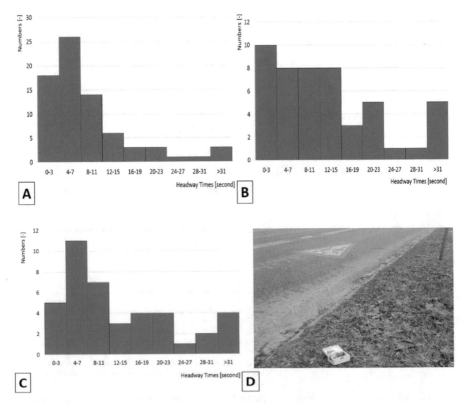

Fig. 11. Characteristics of headway for three cross-sections in one street: (a–c). measurement cross-section 1, 2, 3, d/example of detector positioning

The characteristics illustrated in Fig. 11(a–c) reveal significant differences in the values of headway in respective measurement cross-sections. They mainly result from the changing traffic organisation in this street, but also from the mutual impact between vehicles as well as the fact that the vehicles were approaching pedestrian crossings leading to a school (in a sequence of detectors: 1–2–3, breaking procedure). The distribution of the number of detection entries in consecutive measurement cross-sections has been illustrated in Fig. 12.

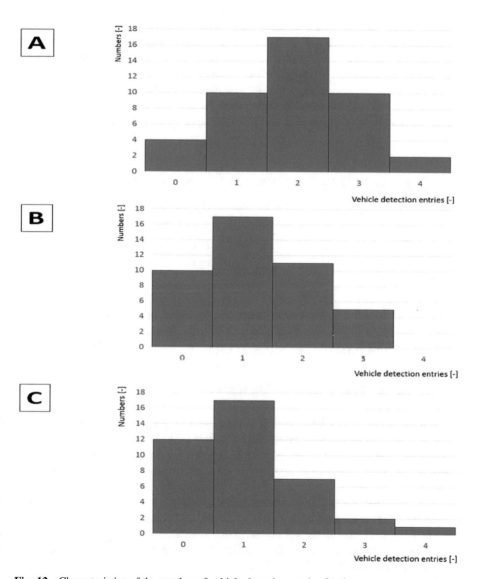

Fig. 12. Characteristics of the number of vehicle detection entries for three cross-sections in one street: (a) cross-section 1-p1, (b) cross-section 2-p2, (c) cross-section 3-p3

The studies discussed above in this paper, conducted in four different (one for two traffic directions) measurement cross-sections, have revealed the following. In a single measurement cross-section (for two opposing traffic directions), differences in the vehicular traffic stream characteristics are relatively small when comparing different traffic directions. However, these difference are becoming increasingly notable when analysing one traffic stream observed in several consecutive measurement cross-sections (for the same traffic direction but before characteristic point). The differences vary depending on the traffic organisation, the stream structure and the traffic conditions. This observation is rather accurately reflected by the randomness of traffic and its capacity to self-regulate in the road network (unlike in railway traffic). No significant differences have been found in terms of the description of traffic characteristics (number of detection entries, headway) compared to those previously elaborated in the literature of the subject. The foregoing pertains to traffic streams of rather low volume. In the case of higher volumes, one may observe e.g. the Erlang distribution. Further research will be conducted in the same street for different traffic volumes, aimed to reveal the dynamics of changes in the characteristics being observed.

Table 1. Results of the χ^2 test for the number of vehicle detection entries.

Cross-section	Calculated value of statistics [-]	Permissible value of statistics [-]
1/E	4.26	7.81
1/W	3.32	7.81
2/p1	5.36	7.81
3/p2	2.26	7.81
4/p3	1.02	7.81

As shown in Table 1, all tested streams are such that the arrival of vehicles has a random character. Which was in accordance with the literature data and the author's expectations. This is appropriate for these objects and for these measurement periods.

4   Analytical Tool - Traffic Simulator

The results of the studies have been uploaded to an original vehicular and pedestrian traffic simulator to analyse the pedestrian crossings (Fig. 13). The simulator uses ore-defined theoretical statistical distributions to describe the given vehicular or pedestrian traffic stream with regard to the pedestrian crossing examined. Instead of theoretical distribution, one can use source data retrieved directly from field surveys (RAW detectors data). Moreover, the simulator makes it possible to modify the parameters of statistical distributions used to simulate vehicular and pedestrian traffic at pedestrian crossings. This allows for the volume of detected vehicles and pedestrians as well as the headway between them to be freely adjusted. The main windows of the simulator has been illustrated in Fig. 13.

The automatic modification of the parameters of statistical distributions describing the vehicular and pedestrian streams makes it possible to simulate the conditions under

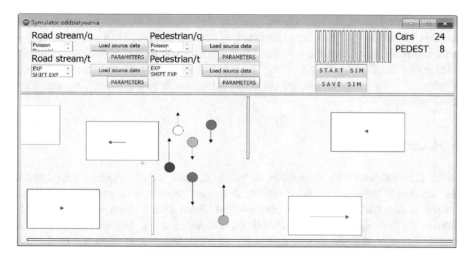

Fig. 13. Traffic simulator used as an analytical tool

which accidents involving pedestrians may take place (congestion and randomness). The foregoing pertains not only to the radical increase in the vehicular and pedestrian traffic volume, but also to simulation of different headway in both these traffic streams. Hence the hypothesis that it is the irregularity of traffic (besides some other significant factors), or rather the sudden and abrupt changes therein, that increases the probability of accidents at pedestrian crossings or in their vicinity. It has been illustrated in Fig. 14.

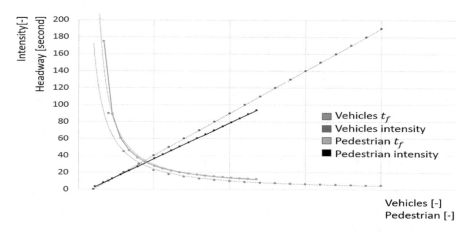

Fig. 14. Linear increase in the intensity of detection entries vs. headway (uniform distribution)

The lines corresponding to the intensity of detection entries delimit specific graph areas where the nature of traffic accidents involving pedestrians is diverse (uniform traffic vs. random traffic).

The basic statistical distributions implemented in the simulator in question have enabled simulation of the characteristics of traffic streams, both of vehicles and pedestrians. Future research will include further distributions, including compound ones, being applied in the program. The relevant simulation results, including those obtained on the basis of compound distributions, will be discussed in subsequent publications.

5   Conclusions

In the pilot studies of traffic streams, neither in the number of detection entries nor in the headway between vehicles and pedestrians were any atypical distributions observed. Moreover, in a single cross-section, these distributions were found to be similar for two different traffic directions. However, some major differences were revealed in the analysis of characteristics of a single street in different road cross-sections (in the area of strong changes in traffic organization). These differences legitimise further research on the variability of characteristics of a traffic stream in different road cross-sections (from a dynamic perspective), setting specific targets for the studies to come. It also requires linking data to OD matrix values. What also requires further investigation is the impact of distance between vehicle or pedestrian presence detectors on values of the characteristics observed. At the same time, the distance between detectors should be chosen in such a manner as to make measurements insensitive to traffic organisation changes (uniform volume stream).

With regard to the simulator developed by the author to analyse traffic at pedestrian crossings, some more methodological remarks, and methods to describe traffic streams and to calculate the probability of traffic disturbance have been provided in other publications, including [8–10]. And although these papers pertain to railway traffic, the traffic modelling methods they describe have also proved useful in road traffic studies.

Some extended research is required in the context of the study. What is the impact of travel behaviour on observed characteristic? A driver which has a delay in journey can significantly affect the local characteristics of the traffic flow. He is driving nervously. This influence should be eliminated. Interesting software tools to study this type of impact are given in the works [13–15]. They require integration with the methodology proposed in this article. This may be possible in future ITS systems via OBU units data.

An very interesting aspect in the context of the study is the impact of road surface condition on observed characteristic (numbers of vehicles, headway). High distortion of road surface can affect the number of vehicles in cross-section and the headway (a large number of small changes duplicates each other). There may be a big shift here? This also requires research in the context of the subject discussed in this article. Pedestrians walk differently on an uneven road. Precisely speaking, this requires a correction of the measured characteristics due to the measured distortion of the road surface [16–20]. As well as these road inequalities can affect directly on traffic events and road fatalities [21].

Currently, the author experiments with a considerably modernized version of the detector. It does not require checking the location with the use of a model detector.

References

1. SEWIK. http://sewik.pl/
2. Drew, D.R.: Traffic Flow Theory and Control. McGraw Hill, New York (1968)
3. Daganzo, C.F.: Fundamentals of Transportation and Traffic Operations. Emerald Group Publishing Limited, Bingley (1997)
4. Haight, F.A: Mathematical Theories of Traffic Flow. Academic Press, Weinheim (1969)
5. Macioszek, E., Lach, D.: Analysis of the results of general traffic measurements in the west pomeranian voivodeship from 2005 to 2015. Sci. J. Sil. Univ. Technol. Ser. Transp. **97**, 93–104 (2017)
6. Macioszek, E., Lach, D.: Analysis of traffic conditions at the Brzezinska and Nowochrzanowska intersection in Myslowice (Silesian Province, Poland). Sci. J. Sil. Univ. Technol. Ser. Transp. **98**, 81–88 (2018)
7. Macioszek, E., Lach, D.: Comparative analysis of the results of general traffic measurements for the Silesian Voivodeship and Poland. Sci. J. Sil. Univ. Technol. Ser. Transp. **100**, 105–113 (2018)
8. Karoń, G., Żochowska, R.: Modelling of expected traffic smoothness in urban transportation systems for ITS solutions. Arch. Transp. **33**(1), 33–45 (2015)
9. Jacyna, M.: Multicriteria evaluation of traffic flow distribution in a multimodal transport corridor, taking into account logistics base service. Arch. Transp. **11**(3–4), 43–66 (1999)
10. Carey, M., Kwieciński, A.: Stochastic approximation to the effects of headways in knockon delays of trains. Transp. Res. Part B **28**, 251–267 (1994)
11. Brilion, W.: Useful estimation procedures for critical gap. Transp. Res. Part A **33**, 161–186 (1999)
12. Daganzo, C.: Estimation of gap acceptance parameters within and across the population from direct roadsides observation. Transp. Res. Part B **15**, 1–5 (1998)
13. Sierpiński, G.: Distance and frequency of travels made with selected means of transport - a case study for the Upper Silesian Conurbation. In: Sierpiński, G. (ed.) Intelligent Transport Systems and Travel Behaviour. AISC, vol. 505, pp. 75–85. Springer, Cham (2017)
14. Sierpiński, G.: Technologically advanced and responsible travel planning assisted by GT planner. In: Macioszek, E., Sierpiński, G. (eds.) Contemporary Challenges of Transport Systems and Traffic Engineering. LNNS, vol. 2, pp. 65–77. Springer, Cham (2017)
15. Sierpiński, G., Staniek, M.: Education by access to visual information - methodology of moulding behaviour based on international research project experiences. In: Gómez, L., Chova, A., López Martínez, I., Torres, C. (eds.) 9th International Conference of Education, Research And Innovation Proceedings, pp. 6724–6729. IATED Academy, Seville (2016)
16. Celiński, I.: Mobile diagnostics of vehicles as a means to examine and define speed limits in a road. Diagnostics **18**(1), 67–72 (2017)
17. Staniek, M., Czech, P.: Self-correcting neural network in road pavement diagnostics. Autom. Constr. **96**, 75–87 (2018)
18. Staniek, M.: Detection of cracks in asphalt pavement during road inspection processes. Sci. J. Sil. Univ. Technol. Ser. Transp. **96**, 175–184 (2017)
19. Burdzik, R., Celiński, I., Czech, P.: Optimization of transportation of unexploded ordnance, explosives and hazardous substances - vibration issues. Vibroeng. Procedia **10**, 382–386 (2016)

20. Staniek, M.: Moulding of Travelling behaviour patterns entailing the condition of road infrastructure. In: Macioszek, E., Sierpiński, G. (eds.) Contemporary Challenges of Transport Systems and Traffic Engineering. LNNS, vol. 2, pp. 181–191. Springer, Cham (2017)
21. Celiński, I., Sierpiński, G.: Availability of data regarding traffic safety and existing infrastructure as a support for optimization of the emergency system. Traffic Eng. **1**, 36–40 (2018)

Improving Chosen Elements of Energy Efficiency in Public Transport Through the Use of Supercapacitors

Grzegorz Krawczyk$^{(\boxtimes)}$ and Jerzy Wojciechowski

Faculty of Transport and Electrical, Kazimierz Pulaski University of Technology
and Humanities in Radom, Radom, Poland
{g.krawczyk, j.wojciechowski}@uthrad.pl

Abstract. This article presents an influence of using an energy storage device in order to reduce the oscillation of voltage and power of the underground traction substation and to limit the peak value of power consumed from it. Using the accumulator has an indirect beneficial influence on the environment because of the efficient energy use. In the work a simulation research of the underground traction substation and a substation equipped with supercapacitors was conducted in the communications peak and beyond it. Simulation studies were performed in toolbox Simscape Electrical formerly called SimPowerSystems and SimElectronics from Simulink products family.

Keywords: Simulation model · Traction substation · Supercapacitor · Energy storage devices · Underground

1 Introduction

Growing pollution of the environment, constant raise of prices of electrical energy and a growing number of vehicles with combustion drive in city agglomerations have become a serious social problem. A solution to these problems, to start with, can be public transport, based on electromobility [1–4]. In big agglomerations the best means of transport of this type is the underground. The underground transport can be characterized with: big capacity, safety, coming on time, energy efficiency and protecting the environment [1, 5, 6].

The nanotechnology development and a growing spectrum of possibilities of use and performance characteristics indicating the advantage of supercapacitors in regards to other energy storage devices can contribute to the fact that the supercapacitors can become, in the nearest future, one of the most popular energy storage devices, both in transport and in power engineering [2, 5, 7, 8].

Among the available storage systems, it seems that the supercapacitors are the most proper ones to be used in the underground power system because of the fast processes of charging and discharging, a high density of power and because of the working conditions in the underground, among others: frequent accelerating and braking and a big power consumption in a short time, etc. There are two types of energy storage devices used: mobile and stationary ones. Installation of an energy storage device inside a

© Springer Nature Switzerland AG 2020
E. Macioszek and G. Sierpiński (Eds.): Modern Traffic Engineering in the System Approach to the Development of Traffic Networks, AISC 1083, pp. 131–142, 2020.
https://doi.org/10.1007/978-3-030-34069-8_11

vehicle or on its top requires additional space and raises the weight of the train, which influences the dynamics of the whole system. An alternative solution allowing the accumulation of energy to the traction network is placing stationary energy storage devices inside the traction substation (preferred in case of the underground). In the above example the size of the accumulator is a secondary issue, which gives more freedom where it comes to choosing the storage device. Using the stationary storage devices in the underground traction substations allows increasing transmission capacity of the power system in emergency conditions and increasing the use of regenerative braking and voltage stabilization in the traction network and, indirectly, decreasing power consumption. The supercapacitor can also constitute an additional energy source (with an appropriate energy capacity), which can be of a vital importance when taking passengers' safety into account during a blackout, thanks to the supercapacitor a traction vehicle can reach the nearest station with a minimum technical speed [2, 5, 7, 8].

2 Simulation Studies of a Traction Substation Not Equipped with an Energy Storage System and with a Supercapacitor

The underground substation with load without an energy storage device and equipped with a supercapacitor was modelled in Simscape Electrical from Simulink products family using the method of differential equation based on solver ode23tb. This method is based on undisclosed trapezoidal rule with a two steps backward equation. The status of the system in a given moment depends on the status from a moment ago and on the current force.

2.1 Simulation Studies of a Traction Substation Not Equipped with an Energy Storage System

The subsystem DC substation (Fig. 1) was modelled as a series circuit of 900 V voltage source that reflects idle voltage of the substation and 23 mΩ of the resistor matching internal resistance of the substation and coil with 15 µH inductance equalling the inductance of the transformer (secondary winding) supplying the rectifier.

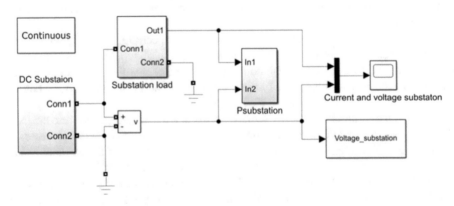

Fig. 1. The underground substation simulation model

Load of the substation connected in series with the DC substation subsystem was modelled as a regulation-current receiver. As input data real values of the underground substation load current were used (The Warsaw metro network has two lines and twenty seven stations forming a rail network of twenty nine kilometers). Values registered on the first metro line during the communications peak and beyond it were taken into account (Fig. 2 frequency of the underground run from 3 to 4 min and beyond the rush hour - from 4 to 5 min). In order to gain a bigger readability of the graphs, only fragments of the hour changes were taken into account (first 20 min).

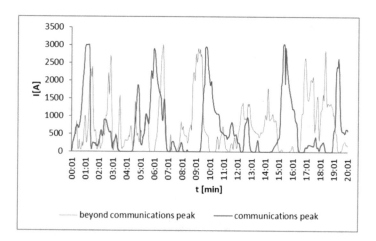

Fig. 2. Changes of instantaneous values of the substation electrical load beyond the communications peak and in the communications peak

Below (Figs. 3, 4 and 5) are the changes received with the substation load beyond the peak and in the peak of the underground run.

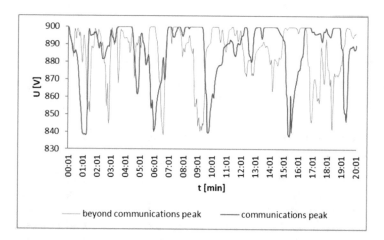

Fig. 3. Changes of instantaneous values of the substation voltage beyond the communications peak and in the communications peak

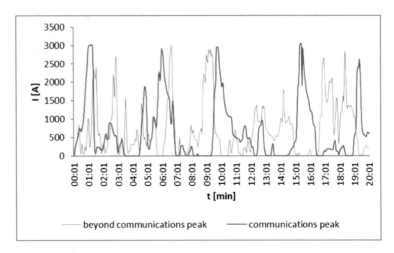

Fig. 4. Changes of instantaneous values of the substation current beyond the communications peak and in the communications peak

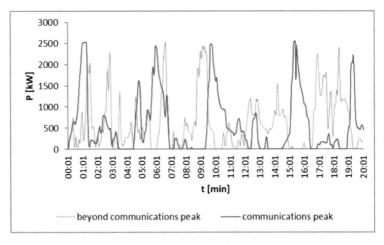

Fig. 5. Changes of instantaneous values of the substation power beyond the communications peak and in the communications peak

2.2 Simulation Studies of a Traction Substation Equipped with a Supercapacitor

Supercapacitors are described with the same laws of physics that regular capacitors are, but have more space for storing bigger load and smaller distances between the electrodes that traditional capacitors have. As a result of this, values of capacity and energy that can be stored in double layer capacitors are increased. Depending on the research approach they can be modelled differently. The energy storage system was modelled as a classical capacitor, without induction terminal (Fig. 6) consisting of $R_1 = 104\ \Omega$

resistance responsible for losses connected in series with a parallel branch consisting in a capacitors C = 90 F and a resistor R_{s-d} = 13 kΩ responsible for self-discharge.

The considered supercapacitor battery consists of 3300 single supercapacitors BCAP3000 of a total weigh of approx. 2 tons (10 parallel branches consisting of 333 double layer capacitors in a series circuit).

A substation model has been equipped with an energy storage system (Fig. 7) with 10 kWh capacity and 900 V voltage.

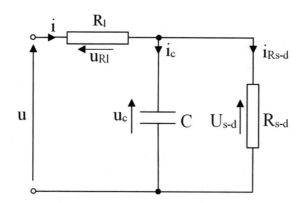

Fig. 6. A model of a double layer capacitor

A system of equations describing the supercapacitor model:

$$\begin{cases} u - \dfrac{u_c R_l}{R_{s-d}} - R_l C \dfrac{du_c}{dt} - u_c = 0 \\ i - \dfrac{u_c}{R_{s-d}} - C \dfrac{du_c}{dt} = 0 \end{cases} \tag{1}$$

Changes of voltage u_c and current i_c:

$$u_c = \left(\frac{u R_{s-d}}{R_l + R_{s-d}} \right) \left(1 - e^{-\left(\frac{t}{\tau}\right)} \right)$$

$$i_c = C \frac{du_c}{dt} = \frac{u}{R_s} e^{-\left(\frac{t}{\tau}\right)} \tag{2}$$

$$\tau = \frac{R_{s-d} R_l C}{R_l + R_{s-d}}$$

Changes of main current i on the other hand, are described by the Eq. (3):

$$i = \frac{u\left[R_l\left(1 - e^{-\left(\frac{t}{\tau}\right)}\right) + (R_l + R_{s-d})e^{-\left(\frac{t}{\tau}\right)}\right]}{R_l(R_l + R_{s-d})} \tag{3}$$

where:

τ - time constant.

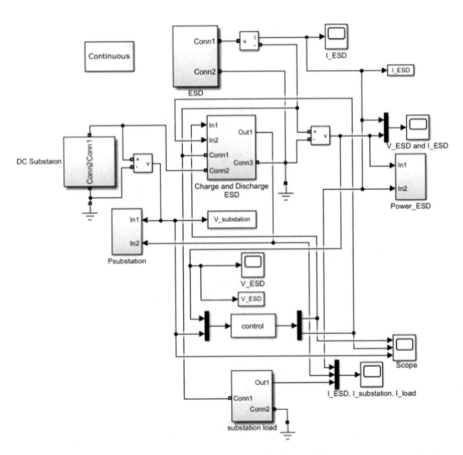

Fig. 7. A simulation model of the underground substation with an energy storage device

After beginning the simulation, one measures an instantaneous voltage on the supercapacitors' clamps. If the value is lower than the rated voltage 900 V, then the supercapacitors will charge to the mentioned value. However, if the temporary voltage of the supercapacitors is equal to the rated voltage (the supercapacitors are fully charged), they will go idle, meaning the supercapacitors will not be charged nor discharged. Because of the above, the supercapacitors are discharged proportionally to the real value of the substation load. In the meantime, if the system is in the status of charging and the following requirements are fulfilled: instantaneous value of the supercapacitors is higher or equal to the rated voltage and the instantaneous value of the substation voltage is lower than the switching voltage - the value of voltage defined by the user (above which the supercapacitors are charged, and below which that are discharged), then the capacitor goes to the discharge mode. Next, when the instantaneous value on the supercapacitors's clamps is lower than the minimal value of the capacitor, and when the instantaneous value of the substation is higher than the switching voltage, then the system will charge the capacitor to the rated voltage. When the instantaneous voltage of the capacitor is higher than the nominal value of the capacitor, and when the instantaneous voltage of the substation is lower than the switching value, then the system will again start the discharge mode. Later, when the conditions are fulfilled: the instantaneous voltage of the capacitor will be lower than its minimal voltage and when the voltage of the substations is higher than the switching voltage, then the system will start charging the supercapacitors one more time. Then, when the instantaneous voltage of the capacitor is higher than the maximum voltage and when the instantaneous voltage of the substation is higher than the switching voltage, then the system will go idle. However, if the instantaneous voltage of the capacitor is lower than the minimal voltage of the capacitor, and the substation voltage is lower than the switching voltage, then the system will go from the discharge mode to the charging mode of the capacitor to the value $V_{ch_{ESD}}$ (voltage value to which the capacitor needs to be charged). The user can define the above value in the range of 900 V–500 V.

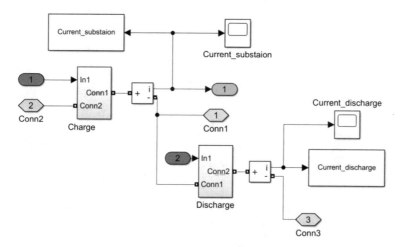

Fig. 8. The actuating system of the supercapacitor

Meanwhile, when the instantaneous value of the capacitor will equal the voltage $V_{ch_{ESD}}$ and the instantaneous voltage of the substation will be lower than the switching voltage, then the system will start to discharge the capacitor. While the instantaneous voltage of the capacitor will be higher or equal than $V_{ch_{ESD}}$ and the instantaneous voltage of the substation will be higher than the switching voltage, then the system will start charging the capacitor to the maximum value of the voltage.

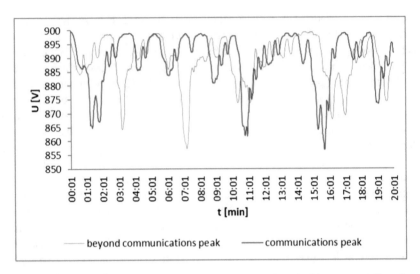

Fig. 9. Changes of instantaneous voltage of a substation equipped with supercapacitors beyond the communications peak and in the communications peak

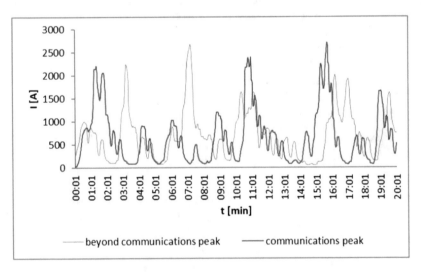

Fig. 10. Changes of instantaneous current of a substation equipped with supercapacitors beyond the communications peak and in the communications peak

The actuating system (Fig. 8) responsible for charging and discharging of the capacitor (Charge and discharge EDLC_ESD) is controlled according to the algorithm placed in the block named "Sterowanie".

Below (Figs. 9, 10, 11, 12, 13 and 14) are the changes (the substation and the capacitor) of voltage, current and power of the substation and the capacitor obtained in the peak energy consumption and beyond it.

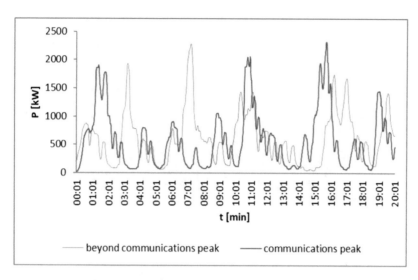

Fig. 11. Changes of instantaneous power of a substation equipped with supercapacitors beyond the communications peak and in the communications peak

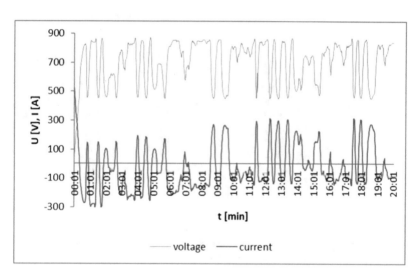

Fig. 12. Changes of instantaneous voltage and current of supercapacitors beyond the communications peak of the underground

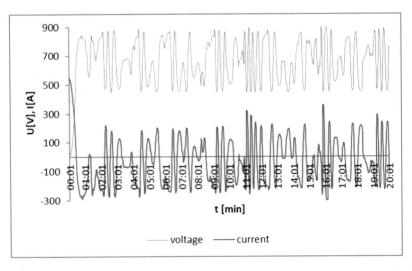

Fig. 13. Changes of instantaneous voltage and current of supercapacitors in the communications peak of the underground

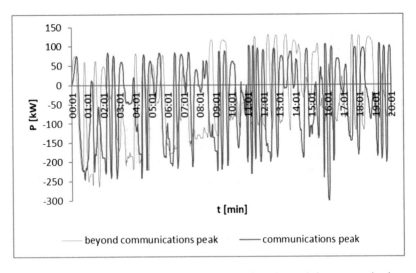

Fig. 14. Changes of instantaneous power of supercapacitors beyond the communications peak and in the communications peak

On the basis of the above changes one can state that using the supercapacitors with proper parameters have caused smoothing of the voltage changes on the substation's clamps, and also limiting current value consumed from the substation, so also the power, in comparison to the model of the substation alone.

3 Conclusions

The authors of the article have expected an improvement of energy efficiency of the electrical traction power supply. As a result of the conducted research and experiments it was possible to demonstrate that the goal has been achieved. The issue of energy efficiency improvement has been connected to the stabilization of voltage conditions in the traction network. The more stable the voltage, the more efficient regenerative braking, which leads to lower energy consumption.

Then, there is a possibility of discharging the capacitor in the moments of bigger power demand and accumulating by the capacitor excess power in the traction network. As a result, voltage oscillation is reduced, which will also lead to limiting overvoltage.

Putting together results acquired from the traction substation equipped with a supercapacitor, the conclusion is that the mean value of voltage of the substation alone is lower in the communications peak by approx. 0.3% (beyond the peak by approx. 0.2%) in comparison to a substation with an energy storage device. The value of the standard deviation is higher in the rush hour by 26.0% (beyond the peak by 22.0%), and coefficient of variation for the option with energy accumulation is lower by 0.4% (beyond the rush hour by 0.3%). While comparing power values of substations with and without a capacitor in the communications peak, power consumption was decreased by around 4.0% (beyond the peak by 16.0%). In the communications peak of traction vehicles' run, the load of the substation is more even than beyond the rush hour. While beyond the peak there are more voltage and current variations. Because of that, the capacitor is used more efficiently beyond the peak (in the analysed case it is approx. 4 times more efficient).

Limiting peak values of power consumption may also influence reducing exceeding power in relation to the ordered capacity for a given substation. Because of that one can contract less capacity than in the substation not equipped with a capacitor, without the fear that it will cause raises in fees due to exceeding power, what is confirmed by the analysis included in this work [2].

Also, the current increase in energy costs causes that all actions rationalizing and limiting energy consumption are wanted from the point of view of economy and a balanced urban development.

On the basis of the above considerations and, among other, works [1, 2] one can state that using energy storage devices in an electrical traction power supply causes an improvement in energy efficiency through decreasing losses in the traction network and limiting ordered capacity for a given substation.

References

1. Xia, H., Chen, H., Yang, Z., Lin, F., Wang, B.: Optimal energy management, location and size for stationary energy storage system in a metro line based on genetic algorithm. Energies **8**(10), 11618–11640 (2015)
2. Szychta, E., Krawczyk, G., Buday, J., Kuchta, J., Michalik, J.: Simulation studies of the underground DC traction substation with and without energy storage device. Commun. Sci. Lett. Univ. Zilina **3**, 26–31 (2013)

3. Kawałkowski, K., Młyńczak, J., Olczykowski, Z., Wojciechowski, J.: A case analysis of electrical energy recovery in public transport. In: Sierpiński, G. (ed.) Advanced Solutions of Transport Systems for Growing Mobility. AISC, vol. 631, pp. 133–143. Springer, Cham (2017)
4. Anthopoulos, L.: Understanding the smart city domain: a literature review. In: Rodríguez-Bolívar, M. (ed.) Transforming City Governments for Successful Smart Cities. Public Administration and Information Technology, vol. 8, pp. 9–21. Springer, Cham (2015)
5. Ciccarelli, F., Iannuzzi, D., Tricoli, P.: Control of metro-trains equipped with onboard supercapacitors for energy saving and reduction of power peak demand. Transp. Res. Part C Emerg. Technol. **24**, 36–49 (2012)
6. Ciccarelli, F., Del Pizzo, A., Iannuzzi, D.: Improvement of energy efficiency in light railway vehicles based on power management control of wayside lithium-ion capacitor storage. IEEE Trans. Power Electron. **29**(1), 275–286 (2014)
7. Dzieliński, A., Sarwas, G., Sierociuk, D.: Comparison and validation of integer and fractional order ultracapacitor models. Adv. Differ. Equ., 1–15 (2011). Springer. https://advance sindifferenceequations.springeropen.com/track/pdf/10.1186/1687-1847-2011-11
8. Buller, S., Karden, E., Kok, D., De Doncker, R.W.: Modeling the dynamic behavior of supercapacitors using impedance spectroscopy. IEEE Trans Ind. **38**(6), 1622–1626 (2002)

Safety Issues in Transport - Human Factor, Applicable Procedures, Modern Technology

Safety Aspects of Female Motorcycling in Slovenia

Peter Jenček[⊠]

Faculty of Maritime Studies and Transport,
University of Ljubljana, Portorož, Slovenia
peter.jencek@fpp.uni-lj.si

Abstract. The use of motorcycles is becoming increasingly popular in Europe due to their numerous advantages. Motorcycling is characterized as a risky activity as the riders are at high risk of getting severely injured or killed in an accident. In the past, motorcycling was often perceived as a male dominated activity. However, the number of female riders is rapidly increasing. Riders' gender and age have been reported to relate to risky behavior in road traffic, often causing motorcyclists' involvement in road accidents. This paper addresses motorcycling attributes and the safety of Slovenian female riders. The analysis performed is based on motor-cycle traffic accidents data and a survey carried out among Slovenian motorcycle riders.

Keywords: Road safety · Motorcycle safety · Motorcycle rider · Female rider · Motorcycling

1 Introduction

According to motorcycle industry estimates, the world motorcycle fleet accounted for 313 million motorcycles in 2008 (almost 80% of which are in Asia) [1]. According to the European Association of Motorcycle Manufacturers, ACEM, there were an estimated 23 million motorcycles in 31 European countries in 2013 [2]. Also according to other sources this share in the total number of vehicles is increasing [3–5].

The use of motorcycles is increasingly popular particularly due to their low cost, high fuel efficiency, comparable speeds to cars, high maneuverability and less space requirement for parking [1]. Motorcycles are used differently in different parts of the world, for recreation (i.e., leisure) mainly in high-income countries, and commercial purposes mainly in low and middle income countries. In Europe, they are very often used for both leisure and commuting (to avoid urban traffic issues, such as traffic jams, queueing, and lack of parking spaces) [6].

The way the motorcycles are used in Europe results in higher market share (over half) of high-powered motorcycles (i.e., over 250 cc) [6]. Over 50% of the motorcycles owned by European motorcycle riders are high-powered motorcycles with a 700 cc or larger engine [2].

Motorcycles belong to a group of vehicles (together with mopeds) referred to as powered two-wheelers (PTW), that are the most dangerous mode of travel in road transport. In 2013, PTWs accounted for more than 286,000 deaths globally, representing

© Springer Nature Switzerland AG 2020
E. Macioszek and G. Sierpiński (Eds.): Modern Traffic Engineering in the System Approach to the Development of Traffic Networks, AISC 1083, pp. 145–162, 2020.
https://doi.org/10.1007/978-3-030-34069-8_12

about 23% of all road traffic deaths in that year [6]. In 2016, about 25,600 people died in accidents on EU roads. PTW fatalities accounted for 17% of those fatalities (18% in 2007) [7].

In Europe, there has been a substantial improvement in motorcycling safety over the last decade. In 2016, more than 3,640 motorcycle riders (drivers and passengers) died in road accidents in the EU, which was about a 38% decrease compared to 2007 (5,821 motorcycle fatalities) [7].

Between 2007 and 2016 the motorcycle fatality rates were reduced in most EU countries, with significant reductions in some. Nevertheless, motorcycling remains one of the modes of transport for which the number of fatalities decreased least between 2007 and 2016. The large majority of motorcycle fatalities in all EU countries are males, with some variation among countries. In 2016, 94% of motorcycle rider fatalities were male and 6% female (half of the female riders were drivers, and the other half were passengers). Between 2007 and 2016, the number of motorcycle rider fatalities decreased for all age groups, except for the older riders (above 50 years). The most frequent causal factors in motorcycle accidents in the EU is excessive speed (too high for the riding conditions, above the speed limit or speed unexpected by other road users), followed by the general category "timing" (consisting of no action, premature action, and late action) [7].

Motorcyclists are at high risk of being involved in a fatal accident or one in which they incur severe injuries. Compared to car drivers, the risk of European motorcyclists of being killed in an accident is from 5–25 times higher in relation to the kilometers driven [8].

The human, vehicle, and environmental factor are considered to have the greatest contribution to the overall outcome of motorcycle accidents, the human being the primary contributing factor for most. Human related factors are found to be the primary accident contributing factor in almost 88% of all cases, thus indicating that the vehicle operators (including motorcycle riders) are largely responsible for accident causation [9].

2 Female Motorcycling

Motorcycling is often characterized as a risky leisure activity that requires physical strength and skill and is dominated by males. In the past, female non-participation in motorcycling (not active as riders, being pillion passengers only) was related to significant physical and monetary risks (as it is considered a costly activity) as well as social risk and prevailing gender norms along with the broader societal image of women's roles in general. However today, in motorcycling, as in many other similarly challenging outdoor recreation activities, the number of female participants is rapidly increasing [10].

In Europe, more men than women and more young than older people ride a motorcycle [8]. In general, the European population of female motorcycle riders is rather small, although it varies from country to country (for example, in France, their share increased from less than 5% in 2001 [11] to 10% in 2012 [12], whereas in Germany their share in 2001 was almost 15% [11]. According to the 2019 Motorcycle Industry Council owner survey, the female motorcycle ownership in the USA increased from 9% in 2010 to 19% in 2018, while even greater female ownership has been identified within younger generations - 22% among generation X (those born from 1961 to 1979) and 26% among generation Y (those born from 1980 to 1999) [13].

Together with the number of female riders, the rate of their involvement in injury and fatal accidents has greatly increased in the last decade. Gender differences are very pronounced in motorcycle crashes, both in terms of number and severity. The relationship between gender and motorcycle riders' motivations and behaviors have been confirmed in several studies [12].

Gender is, together with age and experience, related to riders' attitudes and risk taking behaviors that increase the rider's risk of being involved in a motorcycle accident [14].

As the number of female riders is increasing, it is important to focus more on gender related motorcycling safety research.

3   Motorcycling Safety in Slovenia

3.1   Increase in Motorcycle Fleet

The size of the motorcycle fleet and its growth (changes in the number of registered motorcycles per year), as well as the traveled kilometers per year, are important road traffic system attributes to be considered when analyzing the development of motorcycling and its safety.

The number of registered motorcycles in Slovenia increased almost six times (563%) between 2002 and 2018, from 11,930 to 67,145 vehicles. In the same period, the number of registered cars increased by 27.8%, and the number of all registered motor vehicles increased by 40%. A significant increase in the size of the registered motorcycle fleet occurred from 2006–2010, and again from 2015 onwards (Fig. 1).

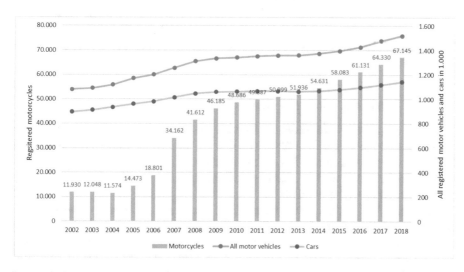

Fig. 1. Number of registered motorcycles compared to cars and all motor vehicles in Slovenia between 2002 and 2018 (Source: own study based on [15])

The share of motorcycles compared to all registered motor vehicles increased from 1.1% in 2002 to 4.4% in 2018 [15]. As the motorcycle fleet is being analyzed in terms of its impact on traffic safety, it is important to mention that in the last 5 years (since 2013), the share of motorcycles older than 12 years increased disproportionately (Fig. 2).

Together with the number of registered motorcycles, the traveled kilometers per year (in million vehicle kilometers per year) also increased. In the period between 2007 and 2017, there was an increase of 155%, from 75.7 to 117 million veh km/year. The rates attributed to the individual type of road remained almost unchanged during the whole period - regional roads around 60%, main roads 20–25%, motor-ways around 15%, and expressways the remaining percentage.

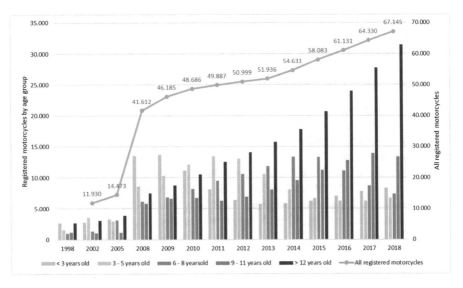

Fig. 2. Age distribution of registered motorcycles in Slovenia between 1998 and 2018 (Source: own study based on [15])

3.2   Development of Motorcycle Safety in Slovenia

The safety of powered two-wheelers (PTW) represents a continuous road safety issue in Slovenia. In the period from 2002 to 2011, the level of motorcycle traffic safety in Slovenia was critical. In 2006, Slovenia ranked worst among the EU countries, with the highest number of PTW rider deaths per billion traveled kilometers (over 350, compared to the EU average of 86) [16].

The risk of motorcycle riders being killed in a traffic accident was more than 50 times higher (EU average 18 times) than the corresponding risk for car drivers. And in the past decade, the ratio of injured motorcycle riders per number of registered motorcycles in Slovenia was significantly higher than the average for EU countries (about 1.5% compared to the EU average of less than 1% in 2008). The statistical data available clearly indicate the higher level of risk related to motorcycle riding in Slovenia compared to the majority of EU countries [17].

During the period 2003–2018, there were 12,428 motorcycle riders (drivers and pillion passengers) involved in police-reported crashes in Slovenia. In the observed period, there were 447 fatalities, 2,629 serious injuries and 6,541 slight injuries among the motorcycle riders in Slovenia, while in 2,838 accidents, no riders were injured [18]. The number of injured motorcyclists per year for each injury category is presented in Fig. 3.

The exact number of motorcycle accidents and the number of riders involved is difficult to determine, mainly due to traffic accident underreporting. Underreporting of traffic fatalities and injuries is a global problem that affects not only low- and middle-income countries. According to the World Health Organization report from 2004, high-income countries are also affected by underreporting (for example, in the United Kingdom, around 36% of road traffic injuries are suggested not to be reported to the police) [19]. We believe that all motorcycle accidents resulting in fatalities and severe injuries in Slovenia are reported to the police. Whereas the motorcycle accidents resulting in slight injuries or no injuries (damage only) are believed to be underreported.

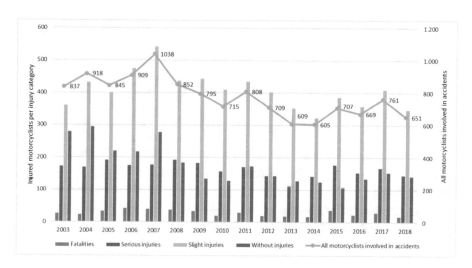

Fig. 3. Number of motorcycle accidents in Slovenia according to injuries suffered between 2003 and 2018 (Source: own study based on [18])

The degree of motorcyclists' safety during the period 2012–2018 has improved slightly, but motorcycle riders are still one of the most vulnerable groups of road users in Slovenia.

The most common causes of traffic accidents involving motorcycles in Slovenia, and caused by riders, are excessive speed and wrong side or direction of driving, followed by a too close (unsafe) vehicle following distance, improper overtaking, failure to give way and vehicle movements [20].

4   Statistical Data on Female Motorcycling Safety in Slovenia Between 2009 and 2018

During the period 2009–2018, there were 7,029 police-reported traffic accidents involving motorcycles in Slovenia. In the observed period, there were 244 fatalities, 1,551 serious injuries, 3,686 slight injuries, and 1,366 accidents with damage only among the motorcyclists [18]. In this period, the female riders' accident involved rate was 5.3%. The rate and number of female and male motorcyclists involved in accidents per year are presented in Fig. 4. There were 375 female riders involved in motorcycling accidents (5.3% of all riders involved in accidents) in the observed period, resulting in 7 fatalities (2.9%), 55 seriously injured (3.5%), 219 slightly injured (5.7%) and 94 (6.9%) accidents without injuries. The rates of female riders' involvement in motorcycle accidents according to the injury suffered, between 2009 and 2018, are presented in Fig. 5.

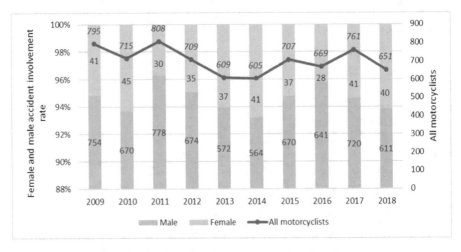

Fig. 4. Rates and number of female and male motorcyclists involved in accidents and number of all motorcycle accidents, 2009–2018 (Source: own study based on [18])

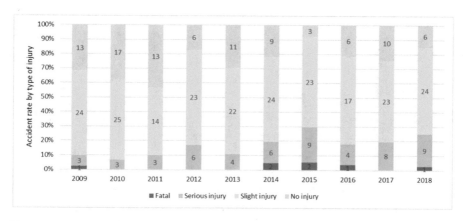

Fig. 5. Rates and number of female riders involved in motorcycle accidents by the injury type, between 2009 and 2018 (Source: own study based on [18])

The age of female riders involved in motorcycle accidents between 2009 and 2018 varied between 15 and 66 (average 35.8 years). The rate and number of female riders of certain age groups involved in accidents vary slightly throughout the observed period. The highest accident involvement rates of female motorcyclists are found for the age groups:

- 16–25: 26.4%, which is higher than the average for rider population (22.4%) as well as male riders (22.2%),
- 36–45: 25.9%, which is higher than the average for rider population (22%) and male riders (21.8%),
- 26–35: 22.1%, which is lower than the average for rider population (24.2%) as well as male riders (24.3%),
- 46–55: 17.6% which is lower than the average for rider population (18.4%) as well as male riders (18.4%).

As for other female riders' age groups, accident involvement rates found are:

- 56–65: 6.9%, which is lower than the average for rider population (9.2%) as well as male riders (9.3%),
- 66–75: 0.3%, which is much lower than the average for rider population (2.6%) and male riders (2.7%),
- 76 and above: there were no female riders involved in accidents.

In the last five years, the female riders aged 60–69 have begun to appear constantly among those involved in accidents. It is interesting to mention that no female rider above 66 was involved in an accident, whereas there were 183 (3%) male riders above age 66 involved in accidents (40 of them aged 75 and above). The rates of riders' involved in motorcycle accidents by age group in the period 2009–2018 are presented in Fig. 6.

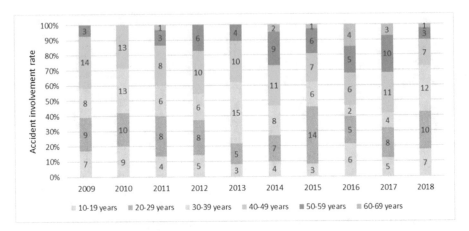

Fig. 6. Rates and number of female riders involved in motorcycle accidents by age group in period 2009–2018 (Source: own study based on [18])

In the observed period (2009–2018) more than 60% female riders on average (ranging between 46% and 70%) were involved in accidents as the responsible party, with an increasing trend (Fig. 7).

The most common causes of accidents involving female motorcyclists in the observed period are excessive speed (with a rate of around 22.0%; male 30.0%), failure to give way (about 18.0%, male 21.0%), wrong side or direction of driving (about 16%; male 12.0%), too close vehicle following distance (about 12.0%; male 8.0%) and risky overtaking (about 7% for both female and male riders). Comparison of female and male riders' population rates for the three most common causes in the observed period is presented in Fig. 8.

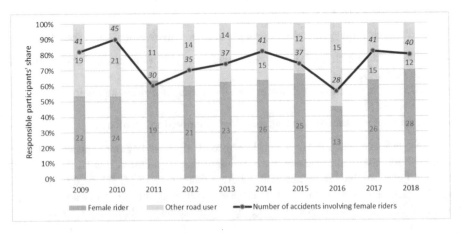

Fig. 7. Share of responsible participants in female riders' accidents (female riders, other road users) in period 2009–2018 (Source: own study based on [18])

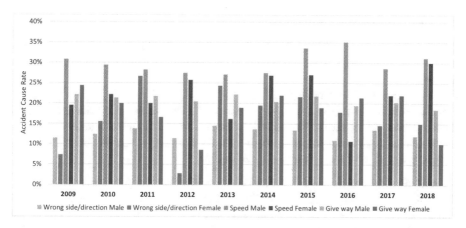

Fig. 8. Comparison of frequency for the three most common causes of riders' accidents for the female and male population in period 2009–2018 (Source: own study based on [18])

The most common types of accidents involving motorcyclists in Slovenia from 2009–2018 were vehicle fall or tilt (24.6%), side-impact collision (23.2%), head-on collision (11.0%), rear-end collision (8.9%) and sideswipe accident (7.1%). These were also the most common types of motorcycle crashes involving female riders with the following rates - vehicle fall or tilt (26.7%), side-impact collision (18.4%), rear-end collision (9.6%), head-on collision and sideswipe accident (9.0% each).

5   Survey on Female Motorcycling Safety in Slovenia

5.1   The Survey

At the Faculty of Maritime Studies and Transport of the University of Ljubljana, there is an ongoing study on the characteristics of motorcycling in Slovenia. The aim of the survey is to collect information about the Slovenian motorcycling community while focusing on motorcycling safety aspects. The survey should provide insight into motorcyclists' profiles, riding habits, safety attitudes, and behavior. It should also identify particularities, similarities, and differences among specific rider subgroups (i.e., female riders, older motorcyclists, members of motorcycle clubs).

The survey was organized as a web based and open participation survey, advertised through several web pages (i.e., Slovenian traffic safety agency, Slovenian national automobile association), motorcycle clubs and Facebook groups.

376 motorcyclists participated in the survey, 244 male riders (65.0%) and 132 female riders (35.0%). As the survey is still ongoing, only selected preliminary results are presented in this paper.

5.2   Female Motorcyclists' Profile

The average age of the motorcyclists participating in the survey was 41.7 years (SD = 12), 37 years (SD = 11.4) for female riders and 43 years (SD = 11.9) for male riders.

The majority of the female riders (80.0%) are long term riders; 13.0% of them have not ridden a motorcycle in the last 24 months, and 7% are returning riders after more than a 2-year riding break. The majority of male riders (81.0%) are long term riders, 3.0% of them have not ridden a motorcycle in the last 24 months, and 16.0% are returning riders after more than 2-year riding break.

Riders' Motivation for Owning and Riding a Motorcycle. The analysis of motives for owning and riding a motorcycle shows that for both male and female riders, the most important motives are: liking motorcycles, the consequent feeling of freedom and independence, spending spare time and traveling. However, there are two interesting features regarding female riders - the "feeling of freedom and independence" and "avoiding dense traffic and delays" appear to be a less important riding motive, while "adrenaline excitement due to (fast) riding" is a more important riding motive for female riders compared to male riders (Fig. 9).

62.0% of female motorcyclists are season-only riders, from March to November (male 43.0%); 33.0% ride mostly during the motorcycling season (male 44.0%), and only 5.0% ride their bike all year round (male 13.0%) (Fig. 10).

In terms of motorcycle usage for leisure riding (i.e., recreational purposes) during the motorcycling season, the survey results show that 18.0% of female riders ride more than 3 times per week (male 27.0%), almost half (48.0%) ride 1–3 times per week (male 41.0%), 27.0% ride 1–3 times per month (male 24.0%), 7.0% ride 5–10 time per season (male 4.0%), while none of them rides only a few times during the whole motorcycling season (male 4.0%). The frequency of motorcycle usage for commuting (i.e., to work, school) during the motorcycling season is slightly different - 15.0% of female riders ride more than 3 times per week (male 12.0%), 17.0% ride 1–3 times per week (male 20.0%), 22.0% ride 1–3 times per month (male 29.0%), 5.0% ride 5–10 time per season (male 10.0%), 24.0% ride only a few times in the whole motorcycling season (male 16.0%), while 17.0% never ride (male 13.0%).

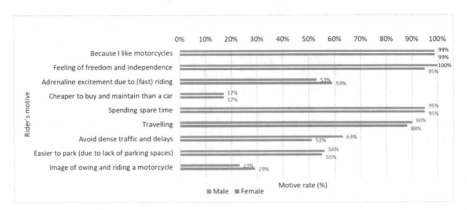

Fig. 9. Slovenian female and male riders' motives for owning and driving a motorcycle (Source: own study based on [21])

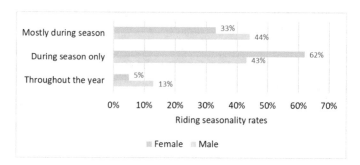

Fig. 10. Motorcycle usage seasonality rates by gender among Slovenian riders (Source: own study based on [21])

Motorcycle Type. The results of the survey among Slovenian motorcycles riders show that the preferred type of motorcycle differs between male and female riders. By far the most common type of motorcycle ridden by the female motorcyclists in Slovenia is "naked" (41.0%), followed by "sport touring" (17.0%), "sport" (14.0%) and "chopper" (14.0%) (Fig. 11).

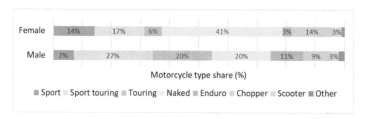

Fig. 11. The preferred types of motorcycles and their shares among Slovenian female and male riders (Source: own study based on [21])

5.3   Female Motorcyclists' Risky Behavior and Accident Involvement

Risky Behavior Related to Riding Speed. Speed is very often identified as a major contributing factor to motorcycle accidents, and motorcyclists often ride at a higher speed than that of car drivers [22].

Exceeding the speed limit is one of the most common traffic offenses among Slovenian riders. The riders' responses in the survey indicate that only a minor part (17.0%) of Slovenian riders were fined for traffic offenses in the last 2 years. However, speeding was the prevailing offense (in 78.0%) among those who committed one, and exceeding the speed limit in urban areas accounted for more than 50.0% of speeding offenses. The results of the survey indicate that views and attitudes related to riding speed may differ considerably between female and male riders. The main findings related to speed when riding a motorcycle can be summarized as presented in Table 1.

Table 1. Views and attitudes of Slovenian riders related to motorcycle riding speed. (Source: own study based on [21, 23]).

View/attitude	Female riders [%]	Male riders [%]
Agree with the statement that motorcycles driving too fast often cause traffic accidents	73.0	85.0
Agree with the statement that cars driving too fast often cause traffic accidents	64.0	85.0
Prefer to ride slowly	37.0	28.0
It is not important if the motorcycle has a high top speed	71.0	70.0
It is important that the motorcycle has fast acceleration	67.0	65.0
Enjoy riding a motorcycle at high speeds	43.0	24.0
Like to corner (ride through the bends) at high speed	29.0	45.0
Feel that driving at 90 km/h on a rural road is too slow	14.0	30.0
Very often/always exceed the speed limit on urban roads	5.0	9.0
Very often/always exceed the speed limit on motorways	11.0	18.0
Very often/always exceed the speed limit on rural roads	21.0	22.0
Often ride so fast into a corner that they scare themselves	9.0	4.0
Often/very often ride so fast into a corner that you feel like you might lose control	18.0	13.0
Often/very often open up the throttle and just "go for it" on rural roads	12.0	24.0
Occasionally/very often exceed the speed limit late at night or in the early hours of the morning	30.0	22.0%

Female Motorcyclists' Group Riding. Although the vast majority (67.8%) of European riders usually ride alone [24], motorcycling is being perceived as an increasingly popular leisure activity that enables riders to meet the like-minded and socialize. Socializing in large groups on public roads, though, can be associated with certain riding safety risks.

Results of the survey indicate that over 70.0% of Slovenian riders are moto-socializing occasionally or on a regular basis. Although Slovenian female riders also prefer to ride alone or with a pillion passenger (67.0% of them), they occasionally do ride in a group (for over 80.0% of female riders group rides amount to less than half of all riding). However, if and when they do ride in a group, it is usually (in 85.0%) a small group of up to 6 riders.

Although group riding can sometimes be related to increased risk of accident involvement (if done in an unsafe way), it can also have a positive impact on motorcyclists' safety. The majority of Slovenian female riders with group riding experience (78.0%) state, that riding in a group is important for them because "it gives them greater fillings of safety, and they can, therefore, be more relaxed during the ride", and because it eliminates the fear related to "not being physically strong enough for lifting the motorcycle from the ground".

Accident and Incident Involvement. The vast majority (80.0%) of female riders have not been involved in a motorcycle traffic accident so far; only a minor part of them (20.0%) had this experience (37.0% of males). And in the past two years, 11.0% of females were involved in car traffic accidents (7.0% males).

The results of the survey indicate that there is a difference in the involvement of Slovenian female and male riders in traffic conflicts (i.e., near accidents) in the past 12 months (Fig. 12). Female riders were less often involved (almost half (43.0%) were not involved at all, 34.0% only 1–2 times and only 5.0% were involved 6 times or more) compared to the more frequent involvement of male riders (only 26.0% were not involved at all, 40.0% only 1–2 times, while 19.0% were involved 6 times or more).

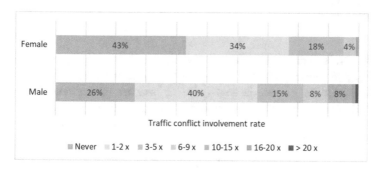

Fig. 12. Female motorcycle riders' involvement in traffic conflicts (frequency in the past 12 months) (Source: own study based on [21])

5.4   Enhancing Female Motorcycling Safety

There are two important physical factors directly affecting riders' safety. Firstly, there are motorcycle safety systems that prevent the occurrence of risky driving situations and thus reduce the risk of motorcycle accidents (i.e., anti-lock brakes, traction and stability control). And then there is motorcycle rider protective equipment that provides physical protection for the rider in case of a fall or collision as well as against natural elements (such as rain, wind, and cold).

Motorcycle Safety Systems. Braking and traction control systems are among the most common safety systems applied to motorcycles. Braking represents an essential aspect of the motorcycle control (in riding speed control and emergency braking in accident avoidance maneuvers) as well as a potential safety issue (i.e. over-braking causing slide-out and fall) [25].

Motorcycles ridden by Slovenian female riders are most commonly equipped with the following safety systems: anti-lock braking system - ABS (71.0%), traction control system - TSC (12.0%), combined braking system - CBS, D-CBS (9.0%), and airbag (2.0%) (Fig. 13). This could indicate that females are riding newer motorcycles (equipped with an ABS system), while males are riding more powerful motorcycles equipped with traction control systems.

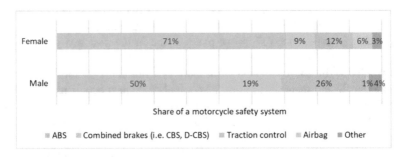

Fig. 13. Slovenian riders' preferred motorcycle safety systems and their shares (Source: own study based on [21])

Motorcycle Rider Personal Protective Equipment. According to the 2015 survey on motorcycling in Europe, the most commonly worn personal protective equipment in Europe are helmets, gloves, jackets (with or without elbow/shoulder protection), motorcycle boots, trousers (with or without hip/knee protection) and back protection [23].

Slovenian riders use protective equipment consistently. Almost all female (91.0%) and male (90.0%) riders always wear some sort of protective equipment. The most commonly used rider protective equipment worn by female riders is protective clothing. Female riders usually wear gloves (91.0%), boots (89.0%), jackets (83.0%) and trousers (68.0%). The details about the wearing rates for other protective equipment are presented in Fig. 14. Slovenian riders were not specifically asked about wearing a helmet, as it has been assumed that the wearing rate is similar to the EU average (>96.0%) [24].

Low motorcycle visibility by other road users, represents a critical motorcycle accident risk factor. An effective way to reduce this risk is to increase the rider's conspicuity by wearing bright or reflective clothes or at least reflective strips or parts of clothes. In the past, a lot of riders (apart from some riders of sport motorcycles) wore mostly dark clothes. Today, riders increasingly wear lighter and reflective colors of clothes. According to the survey, more than half of Slovenian motorcycle riders very often or always wear bright or reflective clothes (43.0% female, 50.0% male), while reflective strips or parts of clothes are commonly worn by 54.0% of the female as well as male riders.

Motorcycle Rider Training. After obtaining their driving license, motorcycle riders can attend voluntary post-license motorcycle rider training courses. The aim of these courses is to increase the riders' risk awareness and technical ability to handle their motorcycles in different (risky) traffic situations, and thus make motorcycling safer and reduce the number of fatalities and seriously injured [26]. There are several types of courses, intended for various segments of riders.

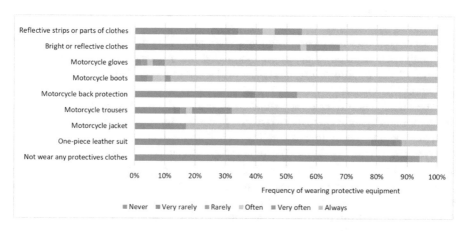

Fig. 14. Wearing rates of Slovenian female riders' preferred protective equipment (Source: own study based on [21])

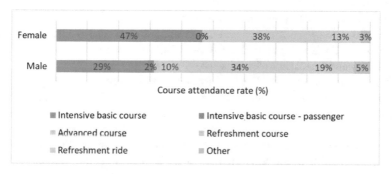

Fig. 15. Types of rider training courses attended by Slovenian female riders and level of participation for each course (Source: own study based on [21])

Slovenian riders are becoming increasingly aware of the benefits of voluntary rider training. Although there are differences in this respect (Fig. 15), between male and female riders - 38.0% of female and 51.0% of male riders, have attended one of the courses. 56.0% of the riders attended the basic (intensive) course, and only 15.0% attended the advanced course, followed by 61.0% of riders that attended a refreshment course at the beginning of the riding season; all of the courses are taking place in a specialized facility (Safe Driving Centre). About one third of the riders attended only a "refreshment ride" on a public road, at the beginning of the motorcycling season.

The three most common reasons for female riders attending the motorcycle training courses are: being aware of the risk related to motorcycle riding in traffic (91.0%), wanting to refresh the driving skills after a winter riding break (82.0%) and being aware of the lack of their own skills for safe riding in traffic (74.0%).

6 Conclusion

This paper addresses the motorcycling attributes and safety of Slovenian female riders. The analysis performed is based on motorcycle traffic accidents data in the period 2009–2018 and preliminary results of an ongoing survey on characteristics of motorcycling in Slovenia.

The human, as a rider or any other road user, is the most important contributing factor for most motorcycle involved traffic accidents. Motorcyclist gender and age have been reported to relate to risky riding behavior in road traffic as well as riders' involvement in road accidents. The motorcyclist population is becoming increasingly diverse as the share of the female as well as older riders is increasing.

Considering riders age is important in analyzing motorcycling safety, as many other key attributes of motorcycling (motorcycle type, size, and usage patterns as well as risky behavior and safety attitudes) are linked to riders' age [8].

The analysis of the statistical data on motorcycle traffic accidents in Slovenia in period 2009–2018, shows that 5.3% of female riders were involved in accidents. Although the results of previous surveys [8] may indicate a much higher proportion (15.0%). Divergence in proportions can be justified by a small number of Slovenian female riders participating in previous surveys.

The rates of female riders' accident involvement with regard to age, are high for the age groups 16–25 (26.4%; male 22.2%), 36–45 (25.9%; male 21.8%), 26–35 (22.1%; male 24.3%) and 46–55 (17.6%; male 18.4%).

There are no research publications available on motorcycling safety in Slovenia, with a particular focus on gender differences. The final results of the survey on characteristics of motorcycling in Slovenia should provide better insight into particularities of different motorcyclist subgroups, in addition to that of female riders, and serve as a valuable contribution to the development of motorcycling safety strategy in Slovenia.

Since the number of Slovenian female riders is increasing, the national motorcycle safety strategy should focus more on their particularities in order to reduce the riding related risks accident involvement and the number of fatalities.

Speeding is a major contributing factor to motorcycle accidents and also the prevailing traffic offense among Slovenian riders while exceeding the speed limit in urban areas accounts for more than half of the speed related offenses.

According to the results of the survey among the Slovenian motorcyclists, the female riders' motives for owning and riding a motorcycle also include "adrenaline excitement due to fast riding", which appears to be more important riding motive for female compared to male riders.

Education and training, as two critical elements of the motorcycling safety strategy, have a significant effect on the rider's behavior and attitudes. As the results of the survey indicate that motorcycle safety related views and attitudes may differ considerably between female and male riders (such as those related to riding speed), these differences should also be taken into consideration when developing the motorcycling safety strategy in Slovenia. The Slovenian motorcycling safety strategy should provide a way to increase the level of awareness of the benefits of voluntary rider education and training among Slovenian riders (as only 38.0% of female and 51.0% of male riders have attended such a course).

The female riders' education could thus increase the level of knowledge and awareness related to motorcycling risks as well as affect their behavior and attitudes (such as those related to respecting the traffic rules and not exceeding the speed limits in particular). The female motorcycle rider training (i.e., rider training takin place in Safe Driving Centers), on the other hand could increase the level of their motorcycle riding skills, particularly in risky traffic situations (i.e., cornering issues, emergency braking, and evasive maneuvers). Such an approach, once it becomes an integral part of the motorcycling safety strategy, would further increase the safety of the female motorcyclists in Slovenia.

References

1. Tiwari, G.: Safety challenges of powered two-wheelers. Int. J. Inj. Control. Saf. Promot. **22**(4), 281–283 (2015)
2. De Bock, H.: Motorcycle Safety and Accidents in Europe. Federation of European Motorcyclists Associations (2016). http://motorcycleminds.org/2016/08/09/motorcycle-safety-and-accidents-in-europe/
3. Macioszek, E., Lach, D.: Analysis of the results of general traffic measurements in the west pomeranian voivodeship from 2005 to 2015. Sci. J. Sil. Univ. Technol. Ser. Transp. **97**, 93–104 (2017)
4. Macioszek, E., Lach, D.: Analysis of Traffic Conditions at the Brzezinska and Nowochrzanowska Intersection in Myslowice (Silesian Province, Poland). Sci. J. Sil. Univ. Technol. Ser. Transp. **98**, 81–88 (2018)
5. Macioszek, E., Lach, D.: Comparative analysis of the results of general traffic measurements for the Silesian Voivodeship and Poland. Sci. J. Sil. Univ. Technol. Ser. Transp. **100**, 105–113 (2018)
6. World Health Organization: Powered two- and three-wheeler safety: a road safety manual for decision-makers and practitioners. World Health Organization, Geneva (2017)
7. European Commission: Traffic Safety Basic Facts (2018). https://ec.europa.eu/transport/road_safety/sites/roadsafety/files/pdf/statistics/dacota/bfs2018_main_figures.pdf
8. SARTRE 4: European road users' risk perception and mobility. The SARTRE 4 survey (2012). http://www.attitudes-roadsafety.eu/home/news/news-single-view/article/sartre4-final-conference/
9. MAIDS: In-depth investigations of accidents involving powered two wheelers. Final Report 2.0. www.maids-study.eu/pdf/MAIDS2.pdf
10. Roster, C.A.: "Girl Power" and participation in macho recreation: the case of female Harley riders. Leis. Sci. **29**(5), 443–461 (2007)
11. Delhaye, A., Marot, L.: PTW Safety and EU Research Work - Review of PTW-related research work from ERSO portal. Annex 21 of the EC/MOVE/C4 project RIDERSCAN (2015). http://www.fema-online.eu/riderscan/IMG/pdf/annex_21.pdf
12. Coquelet, C., Granié, M.A., Griffet, J.: Conformity to gender stereotypes, motives for riding and aberrant behaviors of French motorcycle riders. J. Risk Res. **22**(8), 1078–1089 (2019)
13. Motorcycle Industry Council. https://www.mic.org/#/newsArticle

14. Vlahogianni, E.I., Yannis, G., Golias, J.G.: Overview of critical risk factors in Power-Two-Wheeler Safety. Accid. Anal. Prev. **49**, 12–22 (2012)
15. Statistical Office of the Republic of Slovenia. https://www.stat.si/StatWeb/en/News/Index/8124
16. Šraml, M., Tollazzi, T., Renčelj, M.: Traffic safety analysis of powered two-wheelers (PTWs) in Slovenia. Accid. Anal. Prev. **49**, 36–43 (2012)
17. Topolšek, D., Dragan, D.: Relationships between the motorcyclists' behavioural perception and their actual behavior. Transport **33**(1), 151–164 (2018)
18. The Republic of Slovenia, Ministry of the Interior, Police. https://www.policija.si/eng/
19. World Health Organization: World Report on Road Traffic Injury Prevention. World Health Organization, Geneva (2004)
20. Slovenian Traffic Safety Agency: Pregled stanja varnosti voznikov enoslednih motornih vozil (2019). https://www.avp-rs.si/en/slovenian-traffic-safety-agency/
21. Jenček, P., Hočevar, M.M., Knez, B.: Raziskava o slovenskih motoristih in njihovem udeleževanju v cestnem prometu. Neobjavljena študija. Univerza v Ljubljani, Fakulteta za pomorstvo in promet, Portoroz (2019)
22. Watson, B., Tunnicliff, D., White, W., Schonfeld, C., Wishart, D.: Psychological and social factors influencing motorcycle rider intentions and behaviour. Australian Transport Safety Bureau, Canberra (2007)
23. Sexton, B.F., Baughan, C.J., Elliott, M.A., Maycock, G.: The accident risk of motorcyclists. TRL Report TRL607 (2004). https://trl.co.uk/sites/default/files/TRL607%282%29.pdf
24. Delhaye, A., Marot, L.: The Motorcycling Community in Europe. Deliverable 9 of the EC/MOVE/C4 project RIDERSCAN (2015). https://ec.europa.eu/transport/road_safety/sites/roadsafety/files/pdf/projects_sources/riderscan_d9.pdf
25. Delhaye, A., Marot, L.: Data Collection & Statistics. Deliverable 2 of the EC/MOVE/C4 project RIDERSCAN (2015). https://ec.europa.eu/transport/road_safety/sites/roadsafety/files/pdf/projects_sources/riderscan_d2.pdf
26. Hardy, E.: SMC Survey of Motorcyclists and their views on Advanced Training (2011). http://www.righttoride.eu/virtuallibrary/statistics/SurveySMCfullreport2011.pdf

Non-Technical Aspects of Safety in Scooter-Sharing System in Wroclaw

Agnieszka Tubis, Mateusz Rydlewski[✉], and Emilia Skupień

Wroclaw University of Science and Technology, Wroclaw, Poland
{agnieszka.tubis,mateusz.rydlewski,
emilia.skupien}@pwr.edu.pl

Abstract. Solutions based on the idea of access economy is becoming more and more popular in urban transportation in Polish agglomerations. One of the latest solutions in this system is the sharing of vehicles under scooter-sharing system. Therefore, the purpose of the article is to present the results of research carried out in the area of safety of using shared scooters in the first 3 months of the system implementation in Wrocław. The researchers used direct observation combined with the independent use of vehicles to investigate the issue. The collected experience from the research stage one, became the basis for the preparation of following element of research - an interview based on a survey. It was aimed at examining opinions on the functionality of individual components of the entire rental system among users of the scooter-sharing system. Analyzing the scooter-sharing system in the early phase of its use in Wrocław, 4 areas of existing threats were identified. These hazards can lead to undesirable events that threaten the safety of the user of the scooter or other road users. In the preparation for use phase, aspects related to system registration (including financial settlement) and applicable legal regulations were considered material. In the use phase, the user's behavior, in particular their compliance with the applicable rules, has an impact on operational safety. In the phase of ending the use of the vehicle, the way of leaving the scooter by its current user is of particular importance for the safety of all traffic users. The mobile application, supporting users in vehicle operation, also has a significant impact on the system's safety.

Keywords: Scooter-sharing · Safety · Access economy

1 Introduction

In recent years, a significant progress in the development of information technologies and the growing importance of social networks, in particular in the lives of urban residents has been observed. These phenomena stimulate new economic models based more on cooperation than on competition. These changes illustrate the dynamic development of the so-called sharing economy. This is a trend based on the assumption of better use of resources through sharing, transfer or exchanging products [1]. This concept is defined as a socio-economic system, different from the classic organizational models, built around the division of material and human resources going towards the cross-linking of individuals and society [2]. This system includes the mutual provision

© Springer Nature Switzerland AG 2020
E. Macioszek and G. Sierpiński (Eds.): Modern Traffic Engineering in the System Approach to the Development of Traffic Networks, AISC 1083, pp. 163–173, 2020.
https://doi.org/10.1007/978-3-030-34069-8_13

of services, sharing and co-creation, which allows an increase in the efficiency of the used resources [3].

In urban transport in Polish agglomerations, the solution based on the idea of access economy is becoming more and more popular. One of the latest solutions in this system is the sharing of vehicles under scooter-sharing. Therefore, the purpose of this article is to present the results of research carried out in the area of safety of using scooters in the first 3 months of the system implementation in Wrocław. Achieving such a defined goal requires presentation of the sharing and access economy concept, and then the results of the research will be presented. These studies were focused primarily on aspects related to security issues, but with particular focus on application support, which is an important element of the sharing system.

2   Sharing Economy and Access Economy

2.1   Sharing Economy

The pioneer in the study of the sharing economy phenomenon is Lawrence Lessig, who called in this words the consumption implemented by sharing, exchange and lending one's own resources without transferring ownership of the goods [4]. However, the popularity of this concept began to develop after the publication of the book by Botsman and Rogers *What's Mine Is Yours* [5].

Sharing economy has become popular due to the inhabitants' knowledge about IT techniques and their readiness for interactivity. Generation Y, so-called Millennials are looking at the world differently, and they adapt and perfectly understand peer-to-peer systems. In addition, raising the awareness of residents and striving to limit consumerism means that they need and they are looking for alternatives to their properties, and thus are open to sharing economy [6]. The internet platforms are an important element connecting the sharing economy and the city, allowing to share large amounts of data. Thanks to the possibilities of offering access to large data sets, cities can offer a wide range of services tailored to the expectations of residents. Because of them, sharing mobility can be developed [7].

The number of sharing operators is constantly growing, and it is expected to grow even more rapidly. This is confirmed by a study conducted in 2017 by McKinsey's which shows that among the current users of nontaxi ride-hailing services, 63 percent expect to increase their usage 'a lot' in the next two years, and even more (67 percent) say that they will do the same concerning car sharing [8].

2.2   Access Economy

Eckhardt and Bardhi [9] in their article stated that the sharing economy is not really a 'sharing' economy at all; it is an access economy. According to the authors, sharing sensu stricto is a form of social exchange that occurs between friends and does not bring any profit. If, however, the sharing takes place with the participation of the market, when the company becomes an intermediary between consumers who do not know each other, this practice is based on paid access to the other party's goods and services within a certain time and is in the nature of economic exchange. In this list, clients obtain utility value rather than social value, in which case we are talking about the access economy.

Thus, access-based consumption is usually defined as a transaction that can take place via the market, and in which there is no transfer of ownership [9]. The right of ownership remains with the supplier who bears the related burdens, such as, for example, liability for repairs [10]. The consumer acquires or obtains the right to use the product for a certain period of time, for which he charges a fee when the market mediation takes place [11].

Sharing economy (including, above all, access economy) is increasingly used in various areas of human activity. The areas in which the economy of sharing is the most developed include: transport, storage, tourism, housing, food, media, free time [12]. According to research carried out by PWC, in the year 2025 revenues from consumption based on access will increase to USD 300 billion [13].

Transport is a field in which the sharing economy develops very dynamically, both in the C2C and B2C relations [14]. Urban transport based on the access economy concept is one of the basic instruments by which cities become more intelligent [2]. It is also favoured by the fact that conscious city residents strive to limit consumerism, look for alternatives to ownership and are more open to solutions using sharing economy. Most attention in research of this topic is devoted to *car-sharing*. In the years 2000–2017, 352 scientific publications on this subject were published in reputable journals [13]. At the same time, it is observed that interest in this subject is growing year by year. Also in Poland, more and more people are dealing with these issues in their research. However, most publications are mainly limited to collecting declarations about the behavior and expectations of consumers in Poland regarding car-sharing [13].

3 Scooter-Sharing System Based on Research in Wroclaw

In addition to the ever-growing car-sharing and bike-sharing system in Wroclaw, one can observe the emergence of new solutions for urban transport based on the access economy concepts. A new feature on the Wroclaw market is the scooter-sharing system, which was introduced by the American operator LIME. This operator, being a startup established in January 2017 in the United States, is a pioneer in the studied field. Initially, the company focused on providing its customers with only bicycles, and after about half a year of its activity began work on the rental of electric scooters. After two years of operation, the company makes its vehicles available to users in 92 cities of the United States and in 35 cities outside the US, including Poland.

The sharing idea of a scooter is the - so-called - last mile trip, whose main purpose is to travel directly to the destination of the trip. The scooter from all available vehicles is an unquestionable leader in terms of this possibility. It is possible to reach it almost to the door of the building, which exceeds the possibilities offered by car rental companies, public transport or even city bikes. In addition, an important aspect in this matter are the dimensions of the vehicle, which spatially occupies a bit more space than a pedestrian but much less than a cyclist.

The research conducted along with a group of students from the Wroclaw University of Science and Technology were aimed at assessing the use of scooters in the access economy system and identification of hazards associated with the functioning of this system in urban agglomerations. The research was conducted by a group of 11 people in the period from November 5th, 2018 to January 25th, 2019. These were the first three months of the scooter rental system after its launch in the city. The researchers used direct observation combined with the independent use of vehicles. A group of involved people tested scooters in various conditions of their use. The collected experience has been discussed, and on its basis another element of research - an interview based on a survey was prepared. It was aimed at examining the opinion on the functionality of individual components of the entire rental system among users of the scooter-sharing system. A specially prepared sheet has been published on the Internet. In addition to collecting information directly from users in the form of a questionnaire, some people were involved in tracking users feedback, which were published in social media and on internet forums. This allowed to indicate areas that are a weak side of the rental.

4   Safety Aspects in the Operation of Scooters in the Access Economy System

According to art. 3 of the 'Road Traffic Act' *a participant in traffic and another person on the road are obliged to keep caution or when this act requires extreme caution, avoid any action that might endanger road safety or order, make this movement difficult or disturb the peace or public order and expose anyone to harm.* Analyzing the scooter-sharing system in the early phase of its use in Wroclaw, 4 areas of existing threats were identified: mobile application, user's behaviour, legal regulation, technical condition of the vehicle.

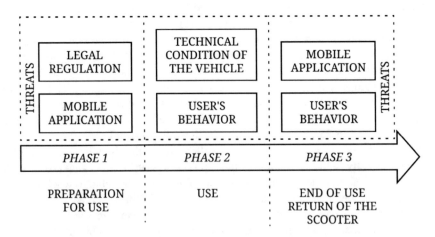

Fig. 1. Three areas of existing threats in scooter sharing system

These hazards can lead to undesirable events that threaten the safety of the user of the scooter or other road users. Each type of threat was assigned to subsequent phases of the scooter use process. The results of the conducted classification are shown in the Fig. 1.

4.1 Preparation for Use Phase

The preparation for use phase begins with user registration in the LIME system. Registering in the electric scooter rental system is much faster than in case of registration in vehicle rental companies. This is due to the lack of user verification. Motor vehicle rental companies verify each new user and check their authorization to drive vehicles available to the system operator. This verification is carried out by sending documents confirming the authorizations held to the operator. In the case of LIME rental, one can use the electric scooter just after downloading the application, accepting the regulations and choosing the payment method. The user of the scooter does not have to confirm in any way that he has skills to navigate the vehicle.

The user has the option to pay for travel in two ways, the first is to provide credit card information from which the fee is charged while using the service (additional pre-authorization is possible). The second form is the ability to top-up your account with 50, 100 or 250 PLN. Direct use during the research period allowed to observe problems with collecting fees consisting in a significant increase in the preauthorisation amount in relation to the actual value of the service. Importantly, funds in such situations can be blocked on one's account for several hours. Emerging problems with financial services cause that some people are afraid to share the data necessary to settle the fee.

The second important aspect affecting the safety of scooter operations in the sharing system is the applicable legal regulations. These regulations apply to both general provisions relating to the use of electric scooters and LIME internal regulations.

There are no regulations in Polish law that regulate the movement of scooters in road traffic [15]. This is mostly due to the fact that Polish law does not qualify scooters for any group of vehicles. Therefore, it is unclear what kind of road infrastructure users can use for scooters. Their movement on the sidewalks is a real threat to pedestrians, because these vehicles are adapted to develop speed up to 25 km/h. On the other hand, the scooter movement on the street significantly influences the safety of its users. This problem is also noticeable outside the country. In France, in which LIME also introduced its scooter sharing system, the Minister of transport undertook to solve in 2019 the problem of the lack of required legal regulations [16]. The statistics on accidents involving scooter users have a significant impact on this. Only in 2017, nearly 300 accidents were registered, as a result of which 5 people died.

The situation is not better if it comes to the rules set by the system provider itself. During the research on the functioning of the LIME sharing system in Wroclaw, the lack of availability of the Polish regulations for the provision of scooter rental services was registered. In the first weeks of the scooter-sharing operation operated by LIME, only the regulations applicable in the United States were accessible, referring to the regulations of individual states. Permanent observations carried out on the company LIME made it possible to notice that the Polish version of the regulations, defining, among other things, the principles of charging or the use of scooters, appeared in the application after 3 months since the service was launched in Wroclaw. The regulations show many inaccuracies and references not related to Polish conditions, such as a fee for using the service in value and currency other than declared and actually charged by the operator.

Irregularities occurring in the phase preceding the use of scooters decide about the risk perceived by potential users of this system. For this reason, people with less willingness to take risk behaviors give up using the system in this form.

4.2   Use Phase

For the safety of the user and other road users in the use phase of the scooter, the technical condition of the vehicle and the user's behavior have a significant impact.

Vehicles delivered in the scooter-sharing system are new, so in most cases their technical condition is satisfactory. Field tests have shown single acts of vandalism, but beyond them the technical condition of vehicles is highly rated. In the surveys conducted, the respondents assessed the following elements of the used vehicles:

- scooter lighting,
- warning bell,
- intuitive acceleration and braking.

Results from of all the above elements presents at Fig. 2.

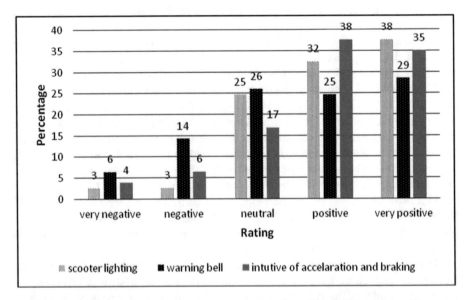

Fig. 2. Respondents' assessment regarding scooter lighting, warning bell and intuitive acceleration and braking

Users of scooters at the moment of rental accept the conditions of its use determined by the supplier. For the sake of the vehicle's operational safety, the four most important principles that should be observed have been identified. Next, on the basis of field observations, the level of compliance with these rules by users was assessed. The assessment was made on a 4-point scale, where: (1) widespread non-compliance with the applicable rules; (2) frequent non-compliance with applicable rules; (3) sporadic non-compliance with applicable rules; (4) strict compliance with applicable rules. The results of the assessment are presented in Table 1.

Table 1. Users' compliance with safety rules.

Safety rule	Level of compliance	Characteristics of observed behaviors
The obligation to ride in a helmet	1	For 3 months of observations no one was found who would have a helmet while driving
Scooter ride only by one person	3	In the evenings and at night people returning from social events use one scooter in pairs - during the observation several such events were noted
Prohibited scooters use on the sidewalk	1	The vast majority of scooters users mainly use sidewalks or bicycle paths if they are available on the route
Prohibition of accelerating from the hill	2	Land falls are a frequent opportunity for users to develop higher speeds than it is possible on flat terrain. Few people slowed down speed before the downhill runs

4.3 End of Use/ Return of the Scooter Phase

The scooter return phase is also an important stage in the use of the scooter-sharing system, which primarily affects the safety of the environment as well as the vehicles. The creators have implemented a system that was to provide them with the ability to control the technical condition of returned vehicles and the method of their transfer for further use. This system consists in the fact that in order to be able to return the vehicle, the current user must do the following: (1) park the scooter; (2) start the application; (3) indicate the willingness to return the vehicle; (4) take a picture of a parked scooter and only then one can finish the rental process. This solution is unfortunately not effective. First of all, the application itself generates many additional problems hindering, and in some cases even preventing the return of the vehicle (more on this subject in point 5). Secondly, the picture taken does not always reflect the actual state of scooter leaving.

Due to the lack of docking stations, the user of the scooter can leave it wherever and whenever he wishes. The carried out observations revealed that users in many cases abandon vehicles in a way that is thoughtless, threatening the lives and health of other road users. Examples of abandoned scooters are shown in the Fig. 3.

The scooters left in the middle of the road or across a poorly lit pavement pose the biggest threat. The oncoming vehicle has a very short response time to the situation, which can lead to accidents. Some users, after returning scooters, consciously place them in places that threaten other people, for example hanging them on a tree branch.

Scooters are also subject to acts of vandalism. They are deliberately spoiled by some users. Examples of damage are shown in the Fig. 3. In the case of visible damage to the scooter, another user can report this through the application. However, if the damage is poorly visible, the next user is exposed to negative events related to the inoperative vehicle that may endanger his or her health, and in extreme cases even life.

Fig. 3. Examples of scooters in LIME system in Wroclaw which have been abandoned or damaged

5   Operation of the Mobile Application

The communication system supporting users of the sharing system is a key element of currently operating solutions. For this reason, the conducted research was also focused on the assessment of the mobile application used by users of the assessed scooter-sharing system.

The most attention was paid to the operation of the mobile application in the phase preceding the use of the scooter and at the time of its return. This is due to the fact that it is phase 1 and 3 that primarily covers the application support, by means of which the service can be provided. The conducted research clearly indicates that users pay attention to the efficiency and reliability of the application. Numerous additional information collected during the survey regarding the failure of the application indicate it as one of the weaknesses of the entire system. This may ultimately help to discourage users from using the service.

The unreliability of the application occurring during the research was noticed by the group of students conducting the research as well as the authors of the article themselves. Irregularities in the operation of the application mainly consisted of problems with the lack of the possibility to rent an available scooter or (in one case) extended by a dozen or so minutes the scooter sentence caused by improper operation of the application. The problems revealed were also noticed by the respondents of the questionnaire who, when evaluating applications, expressed negative opinions about it much more often than it was when assessing other elements of the system. Detailed results are presented in the Fig. 4.

In addition to the unreliability of the application, the method of transmitting messages that seems not to be fully adapted to Polish conditions is important. Observations conducted over the almost 3-month period indicated many areas in which the content presented in the application were in two different languages (Polish and English). While the function consisting in changing the language is very useful, the LIME application had one language version (Polish) with untranslated fragments presented in English. In addition, some of the translations were particularly incomprehensible. The assessment of the intelligibility of text messages in the application by LIME users is presented in the Fig. 4. Observed problems with the application allow to state that it is one of the weakest elements of the entire electric scooter rental system in Wroclaw.

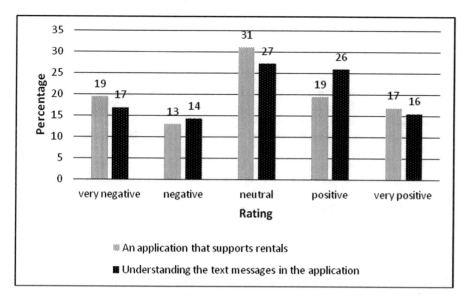

Fig. 4. Respondents' assessment regarding elements of application to rent a scooter in LIME system

6 Summary

The aspects of safety in operation of scooters operating in the access economy described in the article clearly indicate the division into three basic phases that are an integral part of the use of vehicles within the sharing system in the city. These phases are universal for most vehicle rental companies, both passenger cars, scooters, bicycles or scooters described in this article. These phases correspond to the three stages that make up the rental of vehicles and are as follows: preparation for use, use and preparation and passing of the vehicle. Each of these phases is characterized by other areas that can be subjected to various tests.

Phase 2 is the key phase. In principle, it covers the longest period of time in the entire rental process. The use of the vehicle can be considered for a number of reasons. The tests may concern both the reliability of the vehicles being made available, compliance with the provisions of the regulations by users or the determination of users' opinions on individual solutions used in the vehicles provided. A very important element of the city car rental is their high dependence on traffic in the city. When using scooters the key is the impact of other road users on the person using the sharing service.

The research presented in the article covers the period directly after launching the scooter rental system in Wrocław. The adaptation period allowed to indicate the basic problems that users face. Areas shown in the research will be a key element of further research, which due to the nature of the vehicles being made available will be continued from the spring period, in which increased interest in the service is anticipated.

References

1. Burgieł, A.: Wspólna konsumpcja jako alternatywny model spożycia i jej przejawy w zachowaniach konsumentów. In: Kieżel, E. (ed.) Zachowania konsumentów. Procesy unowocześniania konsumpcji. Wolters Kluwer Business, Warszawa (2015)
2. Kauf, S.: Ekonomia współdzielenia (sharing economy) jako narzędzie kreowania smart city. Zeszyty Naukowe Politechniki Śląskiej. Series: Organizacja i Zarządzanie 120, 141–151 (2018)
3. Sokołowski, D., Starzyński, S., Rok, B., Zgiep, Ł.: Ekonomia współpracy w Polsce (2016). http://ekonomiawspolpracy.pl/pobierz-beplatny-raport-wersja-podstawowa/?l
4. Lessig, L.: Remix: Making Art and Commerce Thrive in the Hybrid Economy. Penguin Press, New York (2008)
5. Botsman, R., Rogers, R.: What's Mine is Yours. The Rise of Collaborative Consumption. Harper Business, London (2010)
6. Voznakowa, I., Kauf, S.: Public Transport in the Era of Automatisation and Sharing Economy. Carpatian Logistics Congres, Prague (2018)
7. Martin, E., Shaheen, S.: Impacts of Car2go on Vehicle Ownership, Modal Shift, Vehicle Miles Traveled, and Greenhouse Gas Emissions: An Analysis of Five North American Cities. http://innovativemobility.org/wp-content/uploads/2016/07/Impactsofcar2go_FiveCities_2016.pdf
8. McKinsey. https://www.mckinsey.com/industries/automotive-and-assembly/our-insights/how-shared-mobility-will-change-the-automotive-industry
9. Bardhi, F., Eckhardt, G.M.: Access-based consumption: the case of car sharing. J. Consum. Res. 39(4), 881–898 (2012)
10. Schaefers, T., Lawson, S.J., Kukar-Kinney, M.: How the burdens of ownership promote usage of access- based services. Market. Lett. 27, 569–577 (2016)
11. Durgee, F., O'Connor, G.C.: An exploration into renting as consumption behavior. Psychol. Market. 12(2), 89–104 (1995)

12. Collaborative Finance: The Sharing Economy. http://www.collaborativefinance.org/sharing-economy
13. Tkaczyk, J., Awdziej, M.: Consumer motivations and attitudes toward car-sharing. Res. Paper Wroclaw Univ. Econ. **501**, 165–172 (2017)
14. Koźlak, A.: Sharing-economy as a new socio-economic trend. Res. Paper Wroclaw Univ. Econ. **489**, 171–182 (2017)
15. Public transport news. https://www.transport-publiczny.pl/wiadomosci/hulajnogsharing-juz-w-polsce-60041.html
16. Public transport news. https://www.transport-publiczny.pl/wiadomosci/francja-reguluje-ruch-hulajnogowy-kiedy-polska-60070.html. Last accessed 15 Jan 2019

Assessing the Overtaking Lateral Distance Between Motor Vehicles and Bicycles - Influence on Energy Consumption and Road Safety

Behnam Bahmankhah[✉], Paulo Fernandes, Jose Ferreira,
Jorge Bandeira, Jose Santos, and Margarida C. Coelho

Department of Mechanical Engineering, Centre for Mechanical Technology
and Automation, University of Aveiro, Campus Universitário de Santiago,
Aveiro, Portugal
{behnam.bahmankhah,paulo.fernandes,ferreira.jose,
jorgebandeira,jps,margarida.coelho}@ua.pt

Abstract. The main objective of this paper is to analyse the impacts of the overtaking lateral distance between a bicycle and a motor vehicle (MV) on road safety and energy consumption at two-lane urban roads. An on-board sensor platform was installed on a probe bicycle to measure the overtaking lateral distance and dynamic data. The Bicycle Specific Power (BSP) methodology was used to estimate human required power to ride a bicycle while Vehicle Specific Power (VSP) was used for MVs. The results showed that 50% of overtaking lateral distance were lower than 0.5 m in the peak hours. The BSP and VSP analyses for different values of overtaking lateral distance did not result in any relationship between variables. There was a good fit ($R^2 > 0.67$) between traffic volumes and overtaking lateral distance in the peak hours. On average, the MVs energy consumption in the afternoon was 92% higher than the morning peak periods.

Keywords: Bicycle · BSP · VSP · Traffic · Overtaking lateral distance · Sensor

1 Introduction and Objectives

Cycling offers some important financial, health and social benefits to the users and the environment. Accordingly, cycling is increasing day by day in Europe and in the United States [1, 2]. However, traffic safety concerns could be of high importance for cyclists since they might be more vulnerable to be potentially exposed to injuries in a collision than the driver of a motor vehicle (MV) [3, 4].

In 2016, 2,015 cyclists were killed in road crashes in the European Union (EU28) countries, constituting 8% of all road crashes fatalities [5]. In the same year, 840 cyclists were killed in the United States (US) which accounted for 2.2% of all traffic fatalities [6].

Although bicycle-MV crashes are more severe on rural roads compared to the urban areas [7, 8], the frequency of crashes on urban roads is typically higher. One of the

© Springer Nature Switzerland AG 2020
E. Macioszek and G. Sierpiński (Eds.): Modern Traffic Engineering in the System Approach to the Development of Traffic Networks, AISC 1083, pp. 174–189, 2020.
https://doi.org/10.1007/978-3-030-34069-8_14

main reasons is due to the high speed and manoeuvrability ability of MVs at rural roads [8]. MV speed on rural roads is higher than urban areas, while this high speed can increase safety concerns since it may lead to dangerous overtaking manoeuvres [9].

Other authors also emphasized that MV speed is a fundamental risk factor in cyclist safety mainly when a MV overtaking a bicycle [8, 10]. Ata and Langlois [11] found that the speed of overtaking MVs can affect the lateral distance at urban streets.

Since bicycle size is smaller than MV, it is possible to use more than one bicycle instead of a MV to improve lane use in urban areas. According to the official guidelines from Danish Road Directorate [3], a 2 m wide one-way cycle path has a capacity of 2,000 cyclists while in reality is able to unroll 5,200 cyclists per hour. However, if cyclists use the same lane as MVs, the overtaking lateral distance between bicycle and MV is a key concern regarding cyclists' safety [10, 12]. The overtaking manoeuvrability of drivers [13] can change the behaviour of other MVs and cyclists such as rapidly braking or acceleration. This can represent some safety challenges, especially in narrow lanes and congested traffic situations.

The minimum standard of overtaking lateral distance (the distance between the overtaking MV and the bicycle) in most of the countries is 1.5 m although it is 1 m in some states of the USA [9]. Generally, MVs are required to keep the minimum distance of 1.5 m [8, 9, 12] when passing a bicycle. Overtaking lateral distance is the distance between a MV and a bicycle when the driver is driving straight in the adjacent lane to overtake the bicycle on a road [7, 9].

It is well-recognized that MV overtaking speed is one of the most important parameters affecting MV-bicycle lateral distance, and therefore, the cyclist safety [8, 14]. Debnath et al. [10] measured the overtaking lateral distance between bicycles and MVs based on the speed limit at different zones in the State of Queensland, Australia. They found that when the speed limit is between 70–80 km/h and lower than 40 km/h, the overtaking distance variation comply with the law at curved road sections, and on roads with narrower traffic lanes.

Several studies have shown how the lateral distance variation is influenced by infrastructure design [14–16], MV speed at rural roads [8, 9] and urban roads [17], and driving behaviour [12, 18, 19]. However, there is a lack of research to evaluate the relationship between overtaking lateral distance and specific power considering both MVs and bicycles.

Drivers' decision to keep constant speed or instantaneous decisions to change the speed and subsequently acceleration/deceleration (aggressive driving behaviours) can affect energy consumption, pollutant emissions and safety [15, 20]. Cyclists have more manoeuvrability than MVs but they are more exposed to damage than a MV during a crash [3, 4].

Although riding a bicycle is a simple activity, it requires more human energy for long distances when a conventional bicycle is used. Due to the long-distance travel between origin and destination or road conditions (uphill), a cyclist can feel tired and he/she may not be to use a bicycle [21]. In this context, Mendes et al. [22] developed a methodology to quantify the expended energy of a cyclist using a conventional bicycle which stands for Bicycle Specific Power (BSP). BSP followed a concept widely used to estimate engine load for MV that is the Vehicle Specific Power (VSP) [23].

This regression-based methodology uses dynamic information (speed and acceleration on a second-by-second basis) and topographic conditions (slope) for MV trips.

This paper addressed the impacts of overtaking lateral distance variation between a bicycle and a motor vehicle (MV) on road safety and energy consumption in two urban corridors with variations in cyclist and traffic volumes, and speeds using Global Navigation Satellite System (GNSS) receivers. The main novelty of this paper is the establishment of a relationship between overtaking lateral distance, and BSP, VSP and traffic flow characteristics in different peak hour periods.

The outcome of this work is ultimately to increase the cycling safety at two-lane urban roads by developing a methodology based on the overtaking lateral distance measurements and cyclist/MV energy consumption during the overtaking manoeuvre. Therefore, the specific objectives of this paper are as follows:

- to analyse the driving volatility impact on road safety considering the bicycle and MV overtaking lateral distance variation and acceleration/deceleration variation,
- to assess the relationship between bicycle-MV overtaking lateral distance variation with Vehicle Specific Power (VSP) and Bicycle Specific Power (BSP),
- to assess the impact of traffic volume variation on bicycle-MV overtaking lateral distance variation.

2 Methodology

The methodology of this study relies on field measurements and on-board platform of sensors to measure the overtaking lateral distance between bicycle and MVs (see Fig. 1). Site-specific operations were characterized using videotaping system and manual counting. Concurrently, second-by-second bicycle and vehicle dynamic data were collected using GNSS travel recorders. After that, VSP and BSP were used to compute MV and cyclist energy used during the peak hours, then correlations between overtaking lateral distance and above variables were explored.

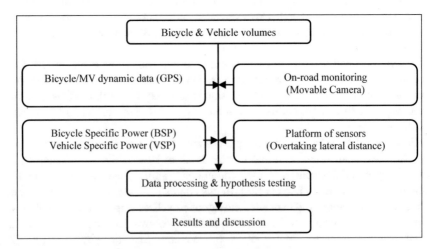

Fig. 1. Methodological framework

2.1 Instrumented Bicycle

The bicycle was instrumented with different sensors and hardware components, as illustrated in Fig. 2. A microcontroller, ESP8266, was used to control and manage the peripheral hardware. The software was developed and compiled on the microcontroller. This component is able to store and process an instruction from the developed software [24].

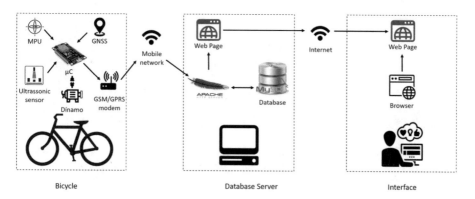

Fig. 2. On-board platform of sensors for enhancing safety of cyclists

To obtain the real-time location coordinates of the vehicle, a GNSS module named GPS-NEO-M8N was used. The system has low power consumption and small dimensions (25 × 35 mm [boars] + 25 × 25 mm [antenna]), and it can receive a signal from various satellite constellations (such as GPS and GLONASS) and follows the NMEA (National Marine Electronics Association) data protocol to communicate with other devices [25].

To track the linear acceleration and angular velocity, the motion-processing unit MPU-6050 was used. This device collects and processes the data from its accelerometers and gyroscopes (one for each axis) and stores the output into memories that can read by the microcontroller. The device also has a temperature sensor [26].

An ultrasonic distance sensor (LV-MaxSonar-EZ1) records the lateral distance of vehicles overtaking by sending ultrasonic waves that are subsequently detected after its reflection in the obstacles. From the time between the sending signal and its echo, the sensor determines the distance to the reflecting object considering the speed of the sound, 340 m/s [27]. All the data obtained by the sensors are collected and pre-processed by the microcontroller. Then it is sent to a GSM/GPRS modem (SIM900) that is responsible for the data transmission to the database server through a mobile network, using TCP/IP messages. To be able to connect to the mobile network, a SIM card is required [28].

This platform of sensors (Fig. 2) can store all dynamic and non-dynamic data in the database server and send it to the end user in real time. The end user can track the cyclist and monitor the bicycle's position and real-time data sent by the sensors. All sensor collected distances that were less than 1.5 m. For purpose of analysis, the results

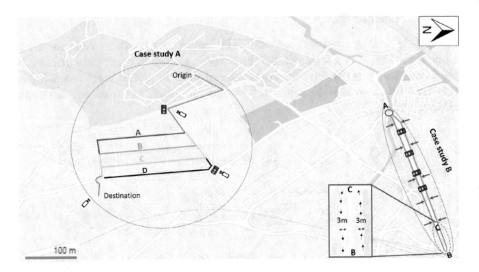

Fig. 3. Layout of the case study with the identification of traffic monitoring points (⬆), case study A (on the left) and case study B (on the right) (Background Source: [31])

Table 1. Data collection specification for each case study.

Case study A		Case study B	
Average (vph) - 8h00–9h30 AM	536	Average (vph) - 8h00–9h30 AM	328
Average (vph) - 5:00–7:00 PM	984	Average (vph) - 5:00–7:00 PM	592
Average (bph) - 8h00–9h30 AM	22	Average (bph) - 8h00–9h30 AM	9
Average (bph) - 5:00–7:00 PM	30	Average (bph) - 5:00–7:00 PM	12
Total road coverage (km)	55	Total road coverage (km)	65
Number of runs	25	Number of runs	23

Note: AM between 8:00–9h30 AM.

are classified into three groups: $x < 0.5$ m, 0.5 m $\leq x < 1.0$ m and 1.0 m $\leq x < 1.5$ m ($x < 1.6$ ft, 1.6 ft $\leq x < 3.3$ ft and 3.3 ft $\leq x < 4.9$ ft).

2.2 Data Collection and Processing

Two case studies with different specifications, such as different average speeds (Sect. 3.2), traffic volumes (Table 1) and road conditions for both bicycle and MV were selected to develop the methodology in the city of Aveiro, Portugal.

The first case study (A) is a corridor with two-lane urban roads at each direction and four intersections with 4 traffic lights (Fig. 3) that is located in the city centre. This corridor was selected since it connects the train station to the city centre, thus representing a relevant trip generator of MVs and bicycles. Case study A has 760 vehicles per hour (vph) and 26 bicycles per hour (bph) at peak hours. The distance between points A and B is 1.1 km, approximately 4 m road width at each direction. Between B and C (~ 250 m) road has only 3 m width with one lane in the travel direction. The second case study (B) is an urban network with four alternative routes (A, B, C and D) between University of Aveiro campus area and one of the city shopping malls (Fig. 3). Traffic movements included two three-leg intersections, one roundabout and four alternative routes. This case study has 460 vph and 10 bph at peak hours. Both case studies A and B are located in a flat terrain.

The sensors, camera and GPS were installed on a conventional bicycle to collect dynamic data and using two different male and female riders between 24 and 37 years old. An equipped light-duty gasoline vehicle with GPS performed several trips along the studied locations where a GNSS device recorded vehicle speed and deceleration/ acceleration rates in a 1-s interval. The minimum number of runs (sample size) was 9 for each direction in each case study based on site-specific traffic signal density (<3 traffic lights/1.6 km) [29]. Thus, 40 GP runs were conducted in this research (20 per site) [29].

Data were collected in four typical weekdays (Tuesday and Wednesday for each case study) during the morning (8 h00–9 h30 AM) and the afternoon (5 h00–7 h00 PM) peak periods. Traffic volumes were counted manually at 5 different points in each direction (Fig. 3) with 15-min intervals for case study A. For case study B, traffic volumes were recorded manually at the entrance of each route by video recording at two signalized intersections and a roundabout near the destination point of the case study (Fig. 3). As in case study A, the traffic volumes were classified in 15-min intervals.

Driving volatility represents the extent of speed and consequently acceleration/ deceleration variations during the MVs movement [30]. The speed and acceleration/ deceleration profiles of bicycles and MVs were extracted to analyse the driving volatility such as sudden or rapid acceleration/deceleration during bicycle-MV interactions.

Critical and extreme variations can occur due to hard acceleration or braking by drivers [30]. After identifying these critical points from data, the reason of these behaviours (peak points) was analysed using video recording.

2.3   VSP and BSP Data Analysis

The selected methodology to estimate the vehicle power consumption variation was based on the concept of VSP that is mathematically defined as follows (Eq. 1) [23, 32]:

$$VSP = v.[1.1\,a + 9.81\ \sin(\arctan{(grade)}) + 0.132] + 0.000302v^3 \tag{1}$$

where:
VSP - vehicle specific power [kilowatt/ton],
v - motor vehicle instantaneous speed [m/s],
a - motor vehicle acceleration/deceleration rates [m/s^2],
$grade$ - terrain gradient [slope].

Each VSP value refers to one of 14 modes for Light Duty Vehicles (LDV) (see Table 2) which in turn are associated with a rate of energy consumption and emissions [22, 33].

Table 2. Binning method for VSP in LDV [23], and BSP for conventional bicycles (Source: [22]).

VSP		BSP	
Range [kW/ton]	Mode	Range [W/kg]	Mode
VSP < −2	1	BSP < −4	< −4
−2 ≤ VSP < 0	2	−4 ≤ BSP < −3	−4
0 ≤ VSP < 1	3	−3 ≤ BSP < −2	−3
1 ≤ VSP < 4	4	−2 ≤ BSP < −1	−2
4 ≤ VSP < 7	5	−1 ≤ BSP < −0	−1
7 ≤ VSP < 10	6	BSP = 0	0
10 ≤ VSP < 13	7	0 ≤ BSP < 1	1
13 ≤ VSP < 16	8	1 ≤ BSP < 2	2
16 ≤ VSP < 19	9	2 ≤ BSP < 3	3
19 ≤ VSP < 23	10	3 ≤ BSP < 4	4
23 ≤ VSP < 28	11	BSP > 4	>4
28 ≤ VSP < 33	12		
33 ≤ VSP < 39	13		
VSP ≥ 39	14		

BSP is estimated second-by-second using the power needed to ride a conventional bicycle, as given by Eq. 2 [25]:

$$BSP = v_{cyclist} \cdot \left[1.01\, a_{cyclist} + 9.81\, \sin(G) + 0.078 \right] + 0.0041 v_{cyclist}^3 \qquad (2)$$

where:
BSP - bicycle specific power [watt/kg],
$v_{cyclist}$ - cyclist instantaneous speed [m/s],
$a_{cyclist}$ - cyclist acceleration/deceleration rates [m/s^2],
G - road grade [slope]

Each BSP value is divided into 11 modes (Table 2) that represent one levels of human energy consumption to ride a conventional bicycle. It should be mentioned that the definition of modes and BSP values varied according to the type of bicycle (e.g. electric bicycle, conventional bicycle [22].

3   Results and Discussion

In this section, the main results from the field measurements are analysed during the bicycle-MV interactions. It proceeds in four sections: First, the overtaking lateral distances are presented (Sect. 3.1) followed by acceleration/deceleration profiles (Sect. 3.2) and resulting VSP-BSP mode distributions (Sect. 3.3). Lastly, the hypotheses are defined and tested (Sect. 3.4).

3.1 Bicycle-MV Overtaking Distance

The extracted data from sensor showed most of the overtaking lateral distance ($\sim 75\%$) occurred in values lower than 1.0 m in both periods (Fig. 4), regardless of the case study. However, the distribution of intervals varied between periods. Regarding case study A, about 56% of bicycle-vehicle overtaking distances were below 0.5 m during afternoon peak but it decreased (in relative terms) in the morning peak (42%). The reason for this result may be due to the differences in traffic volumes between periods (on average, 84% higher in the afternoon peak) that results in less available space for overtaking. Similarly, the results for case study B indicated that about 34% of bicycle-vehicle overtaking distances were below 0.5 m during morning peak hours and increased up to 49% in the afternoon.

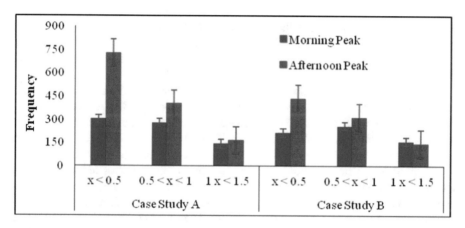

Fig. 4. Bicycle-MV overtaking distance variation (metres) in morning and afternoon peak hours

It is important to emphasise that frequency of the overtaking lateral distance situations for case study B ($\sim 25\%$) is lower than case study A (Fig. 4). It could be due to lower traffic volumes in case study A compared with case study B.

Another explanation for these distances was due to the location where overtaking occurred. For instance, most of these situations occurred near point B (Fig. 3), which has only one circulating lane by direction. Cyclists should avoid riding close to the right edge of the road while at the same time they should care about the overtaking distance from the left side. This situation can increase the risk of a crash in narrow shared lanes.

3.2   Bicycle-MV Acceleration/Deceleration Profile

Bicycles moving at lower speeds than MVs can have more manoeuvrability to use any part of the lane for safety purposes. The average speed values by mode and case study were as follows:

- MVs in case study A - 20 km/h in the morning peak hour and 17 km/h in the afternoon peak hour,
- bicycles in case study A - 13 km/h in the morning peak hour and 10 km/h in the afternoon peak hour,
- MVs in case study B - 28 km/h in the morning peak hour; 22 km/h in the afternoon peak hour,
- bicycles in case study B - 15 km/h in the morning peak hour and 12 km/h in the afternoon peak hour.

The above-mentioned results indicated higher speed values in case study B, regardless of the peak period. This point may be explained by the high volume-to-capacity ratio in case study A (up to 0.65) even though corridor has two lanes in travel direction.

The analysis of the bicycle acceleration/deceleration profiles showed similar profiles within transport mode in the morning and afternoon regardless of the case study. Bearing this in mind, one profile was selected from the morning and afternoon data samples for bicycles and MVs, as shown in Fig. 5. It was found that, regardless of the case study, the range of acceleration/deceleration rates in the morning was higher than in the afternoon. This may be due to the fact that traffic volumes are higher in the afternoon peak periods, resulting thus in more stop-and-go cycles due red signals, pedestrians at crosswalks or yielding to circulating traffic at roundabouts. Other reason behind these peak points of driving volatility is that MVs or cyclists did braking manoeuvres to avoid the crash with those vehicles that were moving from the parking area to the travel lane or because of suddenly opening of the door into the path of the bicycle. There was some evidence that the high bicycle acceleration/deceleration rates were caused by drivers who opened car doors into the path of an approaching cyclist or others who illegally parked MVs at the right-hand side of the road (mainly on the bicycle path).

Riding and driving behaviours of cyclists and drivers were analysed at narrow sections of lanes (250 m from C to B and B to C) in case study A (see Fig. 2). Travel start and stop times from point C to B were extracted from videotapes using GPS data. The results showed that there is no evidence of high acceleration/deceleration rates and MVs behaviour and manoeuvrability were proper at narrow sections of lanes while the most overtakes of less than 0.5 m overtaking lateral distance occurred at these sections of lanes. Cyclists have to pay more attention to left and right sides when the lane is narrow, whereas it seems that drivers also care more about cyclists in these areas.

3.3   VSP and BSP Modes Distribution

VSP and BSP values were calculated against the time spent by MV and bicycle in their modes. Figure 6a, and c represent the distribution of BSP modes for a conventional

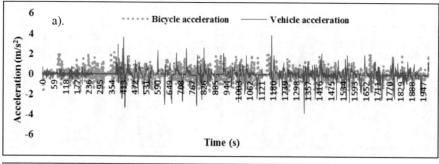

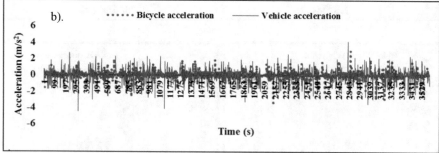

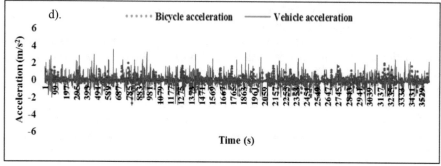

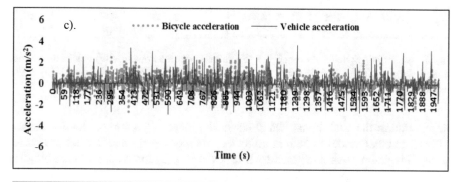

Fig. 5. Bicycle and MV acceleration/deceleration profiles at peak hours: (a) morning case study A; (b) afternoon case study A; (c) Morning case study B; (d) afternoon case study B

a).   b).

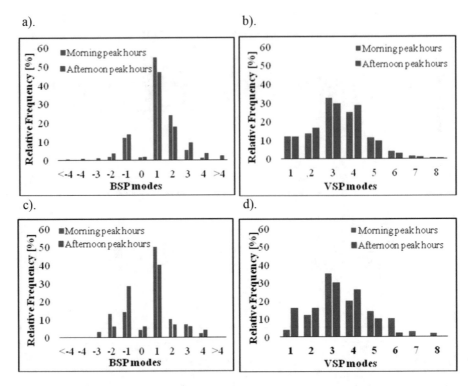

c).   d).

Fig. 6. (a) Bicycle Specific Power (BSP); (b) Vehicle Specific Power (VSP) modes distribution in case study A; (c) Bicycle Specific Power (BSP); (d) Vehicle Specific Power (VSP) modes distribution in case study B, in morning and afternoon peak hour

bicycle during the peak hours. On average, the bicycle spent more time in mode 1 (50%) than other modes in both morning and afternoon periods regardless of the case study. This finding was confirmed by Mendes et al. [22] about conventional bicycles. The higher percentage of mode 1 may be since cyclists do not need high human power to ride a bicycle or may be due to the low speeds of bicycles. Regarding the distribution of the modes, no significant differences were observed between the morning and the afternoon peak hours. To examine the consistency between the morning and afternoon VSP mode distributions, the two-sample Kolmogorov-Smirnov statistical test (K-S test) for the analysis of histograms with 99% confidence level was used for both case studies A and B. The mean of BSP values showed only 7% and 9% difference between morning and afternoon in case studies A and B, respectively.

The results confirmed that on positive BSP modes, the bicycle spent more time (83% and 79% in case study A and B, respectively) compared with the negative modes (17% and 21% in case studies A and B, respectively) in both morning and afternoon peak hours. Figure 6b, and d represent the distribution of VSP modes for MVs during the peak hours. VSP modes distribution in case study A are approximately same in the morning and the afternoon peak hours while the variation of VSP modes is higher in case study B. It can be due to the more space available for the movement in case study

B than A. The average speed of case study B (25.2 km/h) (considering both the morning and afternoon periods) was 34% more than case study A value (18.8 km/h).

MVs spent on average more time in mode 3 (31% and 28% in case study A and B respectively) and mode 4 (27% and 23% in case study A and B respectively) than other modes in both the morning and afternoon peak hours. Mode 3 represents idling and low speed situations while mode 4 represents accelerations at low speeds.

CO_2 emissions were calculated based on the VSP concept [23] for the gasoline MVs in peak hours in order to assess the energy consumption. Regarding the direct correlation between CO_2 emissions and energy consumption, energy consumption was found to be increased by 92% in the afternoon compared with the morning.

The results showed that 17% and 32% of CO_2 emissions in the morning and 16% and 37% in the afternoon were generated in mode 3 (idling or low speed situations) and mode 4 (acceleration at low speeds) in case study A. About case study B, mode 3 and 4 have corresponded to 14% and 27% of CO_2 emissions in the morning and 12% and 16% in the afternoon in case study B. As shown in Fig. 6b–c, the frequency of time distribution for modes 3 and 4 was approximately the same in the morning and the afternoon. The mean of VSP values showed only 4%–6% difference between morning and afternoon.

3.4 Hypothesis Testing

The relationships between overtaking lateral distance and VSP/BSP mode and traffic volume were investigated. The morning overtaking lateral distance values were on average higher than the afternoon period when the traffic volumes were lower. Therefore, the authors decided to evaluate the impact of traffic volume on overtaking lateral distance variation and the following hypothesis was defined:

- overtaking lateral distance variation was expected to have less impact on VSP and BSP modes than traffic volumes.

The results from Fig. 7 seem to confirm above hypothesis. First, no correlation (coefficient of determination - $R^2 < 0.07$ and $R^2 < 0.02$ in case studies A and B, respectively) was found between VSP/BSP and overtaking distance variation, regardless of the time period.

Second, scatter plots indicated that traffic volumes (15-min intervals) and overtaking lateral distance followed a linear trend both in the morning and afternoon peak hours ($R^2 = 0.68$ and $R^2 = 0.73$ in case studies A and B, respectively). Both intercept and slope parameters had p-values lower than 0.05, thus indicating statistical significance. Table 3 summarises the statistical analysis of models separately for the morning and afternoon periods.

Results from Fig. 7 and Table 3 confirmed that the traffic volume variation had a moderate effect on overtaking lateral distance between bicycles and MVs during the peak hours. Regardless of the case studies, it can be concluded that overcoming lateral distance between the bicycle and the MVs decreases with increasing traffic volume.

Since the traffic volumes were collected at different segments of case study A and at different routes of case study B, the results of correlation between overtaking lateral distance and traffic volume variation can be applied to all the corridor (case study A)

a).   b).

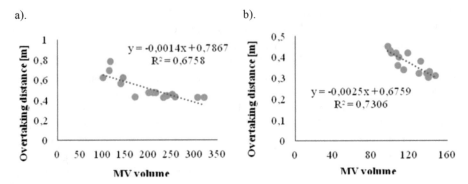

Fig. 7. Correlation between overtaking distance variation and traffic volume variation: (a) Morning and afternoon peak in case study A; (b) Afternoon and afternoon peak hour in case study B

Table 3. Summary of statistical analysis for model coefficients between traffic volumes and overtaking lateral distance variation.

Period		Model parameter	Coefficients	Standard Error	T statistics	p-value
Case study A	Morning peak	Intercept	946.2	199.7	4.7	0.003
		Variable X1	−1565.2	445.8	−3.5	0.013
	Afternoon peak	Intercept	252.0	43.9	5.7	0.004
		Variable X2	−195.6	72.1	−2.7	0.037
Case study B	Morning peak	Intercept	736.21	217.6	3.1	0.011
		Variable X1	−423.3	332.1	−2.2	0.004
	Morning peak	Intercept	317.1	100.5	3.8	0.002
		Variable X2	−1205.4	215.3	−4.1	0.027

and the network (case study B). The linear coefficient model within a 95% confidence level was applied to show the relationship between traffic volumes and overtaking lateral distance. Variable Y represents the overtaking lateral distance while $X1$ and $X2$ represent traffic volumes in the morning and afternoon periods, respectively.

4   Conclusions

This paper represents an evaluation of the impacts of the bicycle-MV overtaking lateral distance on driver and cyclist behaviours, safety and BSP/VSP mode distributions. Field measurements were conducted in a real-world corridor with traffic lights and an urban network with four alternative routes. The analysis was based on overtaking lateral distance measurements extracted from a platform of sensors installed on a conventional bicycle. Measurements were carried out in morning and afternoon peak hours. Bicycle and MV GPS data were also used to characterize road user behaviours.

More than 75% of the total overtaking lateral distances were lower than 1.0 m, and 50% were lower than 0.5 m, thus confirming some issues regarding the cyclist safety. It was found that lowest overtaking lateral distances (<0.5 m) occurred in segments with high traffic volumes segments with resulting lack of road space during the interaction between motor vehicles and cyclists. The analysis of acceleration/deceleration profiles confirmed that bicycles and MVs had similar behaviour in both periods, but the trend of acceleration/deceleration for MVs was higher than bicycles regardless the case studies.

The analysis of relationship for traffic volumes and overtaking lateral distances showed moderate to good fit between these variables ($R^2 = 0.68$ and $R^2 = 0.73$ for case studies A and B respectively). In contrast, no correlation was observed between overtaking lateral distance and BSP/VSP modes.

Although dynamic data used in this paper was stored before processing, one of the main contributions of this paper is the integration of real-time driving volatility information on a platform to alert road users about potential proximity with cyclists, and as result, some crashes.

Future study would consider the impact of age, gender or colours of clothes of cyclists on overtaking distances, and different types of road.

Acknowledgments. The authors acknowledge to the following projects: Centre for Mechanical Technology and Automation Strategic Project UID/EMS/00481/2019-FCT and CENTRO-01-0145-FEDER-022083; @CRUiSE (PTDC/EMS-TRA/0383/2014), funded within Project 9471–Reforçar a Investigação, o Desenvolvimento Tecnológico e a Inovação (Project 9471–RIDTI) and supported by European Community Fund FEDER; MobiWise (P2020 SAICTPAC/0011/2015), co-funded by COMPETE2020, Portugal2020-Operational Program for Competitiveness and Internationalization (POCI), European Union's ERDF (European Regional Development Fund) and FCT - Portuguese Science and Technology Foundation; CISMOB (PGI01611, funded by Interreg Europe Programme); DICA-VE - Driving Information in a Connected and Autonomous Vehicle Environment: Impacts on Safety and Emissions (POCI-01-0145-FEDER-029463), funded by FEDER, through COMPLETE2020-Portuguese Operational Program Competitividade e Internacionalização (POCI), and by National funds (OE), through FCT/MCTES. The authors also acknowledge to Órbita Bikes, which allowed the use the bicycle also Mariana Vilaca and Carlos Sampaio for their cooperation in data collection.

References

1. Pucher, J., Buehler, R., Seinen, M.: Bicycling renaissance in North America? An update and re-appraisal of cycling trends and policies. Transp. Res. Part A Policy Pract. **45**(6), 451–475 (2011)
2. Pucher, J., Buehler, R.: Cycling towards a more sustainable transport future. Transp. Rev. **37**(6), 689–694 (2017)
3. Van Hout, K.: Annex I: Literature search bicycle use and influencing factors in Europe. BYPAD. EIE-programme 05/016 Intelligent Energy Europe. Instituut voor Mobilitet, UHasslt (2008)
4. Götschi, T., Garrard, J., Giles-Corti, B.: Cycling as a part of daily life: a review of health perspectives. Transp. Rev. **36**(1), 45–71 (2016)

5. European Commission: Traffic Safety Basic Facts 2018 - Cyclists. European Road Safety Observatory Documentation. Traffic Safety Basic Facts. European Commission, Brussels (2018)
6. U.S. Department of Transportation: National Highway Traffic Safety Administration, National Center for Statistics and Analysis: Bicyclists and Other Cyclists: 2016 Data. Report No. DOT HS 812 507. National Highway Traffic Safety Administration, U.S. Department of Transportation, Washington (2018)
7. Dozza, M., Schindler, R., Bianchi-Piccinini, G., Karlsson, J.: How do drivers overtake cyclists? Accid. Anal. Prev. **88**, 29–36 (2016)
8. Stone, M., Broughton, J.: Getting off your bike: cycling accidents in Great Britain in 1990–1999. Accid. Anal. Prev. **35**(4), 549–556 (2003)
9. Llorca, C., Angel-Domenech, A., Agustin-Gomez, F., Garcia, A.: Motor vehicles overtaking cyclists on two-lane rural roads: analysis on speed and lateral clearance. Saf. Sci. **92**, 302–310 (2017)
10. Debnath, A.K., Haworth, N., Schramm, A., Heesch, K.C., Somoray, K.: Factors influencing noncompliance with bicycle passing distance laws. Accid. Anal. Prev. **115**, 137–142 (2018)
11. Ata, M.K., Langlois, R.G.: Factoring cycling in transportation infrastructure: design considerations based on risk exposure. ITE J. **81**(8), 49–53 (2011)
12. Feng, F., Bao, S., Hampshire, R.C., Delp, M.: Drivers overtaking bicyclists - an examination using naturalistic driving data. Accid. Anal. Prev. **115**, 98–109 (2018)
13. Chapman, J., Noyce, D.: Observations of driver behavior during overtaking of bicycles on rural roads. Transp. Res. Rec. J. Transp. Res. Board **2321**, 38–45 (2012)
14. Shackel, S.C., Parkin, J.: Influence of road markings, lane widths and driver behaviour on proximity and speed of vehicles overtaking cyclists. Accid. Anal. Prev. **73**, 100–108 (2014)
15. Wang, X., Khattak, A.J., Liu, J., Masghata-Amoli, G., Son, S.: What is the level of volatility in instantaneous driving decisions? Transp. Res. Part C Emerg. Technol. **58**, 413–427 (2015)
16. Mehta, K., Mehran, B., Hellinga, B.: Evaluation of the passing behavior of motorized vehicles when overtaking bicycles on urban arterial roadways. Transp. Res. Rec. J. Transp. Res. Board **2520**, 8–17 (2015)
17. Chuang, K.H., Hsu, C.C., Lai, C., Doong, J.L., Jeng, M.C.: The use of a quasi-naturalistic riding method to investigate bicyclists' behaviors when motorists pass. Accid. Anal. Prev. **56**, 32–41 (2013)
18. Kay, J.J., Savolainen, P.T., Gates, T.J., Datta, T.K.: Driver behavior during bicycle passing maneuvers in response to a Share the Road sign treatment. Accid. Anal. Prev. **70**, 92–99 (2014)
19. Duthie, J., Brady, J., Mills, A., Machemehl, R.: Effects of on-street bicycle facility configuration on bicyclist and motorist behavior. Transp. Res. Rec. J. Transp. Res. Board **2190**, 37–44 (2010)
20. Liu, J., Khattak, A.J., Wang, X.: A comparative study of driving performance in metropolitan regions using large-scale vehicle trajectory data: Implications for sustainable cities. Int. J. Sustain. Transp. **1**(3), 170–185 (2017)
21. Dill, J., Rose, G.: E-bikes and transportation policy: insights from early adopters. Transp. Res. Rec. J. Transp. Res. Board **2314**(1), 1–6 (2012)
22. Mendes, M., Duarte, G., Baptista, P.: Introducing specific power to bicycles and motorcycles: application to electric mobility. Transp. Res. Part C Emerg. Technol. **51**, 120–135 (2015)
23. Frey, H.C., Unal, A., Chen, J., Li, S., Xuan, C.: Methodology for developing modal emission rates for EPA's multi-scale motor vehicle & equipment emission system. Environmental Protection Agency, Raleigh (2002)

24. User Manual V1.2: ESP8266 NodeMCU WiFi Devkit. Hanson Technol. http://www.handsontec.com/pdf_learn/esp8266-V10.pdf
25. U-BLOX: NEO-M8, u-blox concurrent GNSS modules - Data Sheet u-blox. https://www.u-blox.com/sites/default/files/NEO-M8_DataSheet_(UBX-13003366)
26. InvenSense: MPU-6000 and MPU-6050 Product Specification. InvenSense Inc. www.invensense.com
27. MaxBotix: LV-MaxSonar ® -EZTM Series. MaxBotix Inc. www.maxbotix.com
28. SIMCom: SIM900: Hardware Design. Smart Mach Smart Decis. Shangai: SIMCom Wireless Solutions Ltd. http://imall.iteadstdio.com/IM120417009_IComSat/DOC_SIM900_HardwareDesign_V2.00.pdf
29. Li, S., Zhu, K., Gelder, B.H., Nagle, J., Tuttle, C.: Reconsideration of sample size requirements for field traffic data collection with global positioning system devices. Transp. Res. Rec. J. Transp. Res. Board **1804**, 17–22 (2002)
30. Khattak, A.J., Wali, B.: Analysis of volatility in driving regimes extracted from basic safety messages transmitted between connected vehicles. Transp. Res. Part C Emerg. Technol. **84**, 48–73 (2017)
31. Bing Maps. https://www.bing.com/maps
32. Anya, A.R., Rouphail, N.M., Frey, H.C., Liu, B.: Method and case study for quantifying local emissions impacts of transportation improvement project involving road realignment and conversion to multilane roundabout. In: Transportation Research Board 92nd Annual Meeting Compendium of Papers, pp. 1–22. Transportation Research Board, Washington (2013)
33. Coelho, M.C., Frey, H.C., Rouphail, N.M., Zhai, H., Pelkmans, L.: Assessing methods for comparing emissions from gasoline and diesel light-duty vehicles based on microscale measurements. Transp. Res. Part D Transp. Environ. **14**(2), 91–99 (2009)

Analysis of the Relationship Between Traffic Accidents with Human and Physical Factors in Iraq

Firas Alrawi[✉] and Amna Ali

Urban and Regional Planning Center, University of Baghdad, Baghdad, Iraq
dr.firas@uobaghdad.edu.iq, dr.amna.s.ali@iurp.
uobaghdad.edu.iq

Abstract. Traffic accidents are one of the principal causes of the loss of abundant human and natural resources and destruction of infrastructure in all countries, especially developing one. This research aims to measure the relationship between the range of variables related to the types of traffic accidents (crash, rollover, run over, and others), with the most common causes that lead to these accidents (road, vehicle, driver, and pedestrian) in Iraq. This relationship based on the hypothesis that the variance of these accidents is the result of the variation of their causes, this relationship was tested by using the multiple regression analysis model for tow periods of study time (1979–2000) and (2010–2017). This test demonstrated a strong correlation between traffic accidents and the causes of accidents. The rate of correlation of variables in the general estimated models between 95% and 100%, which mean there is a strong positive correlation between traffic accidents and road users, driver and pedestrian (human factors), and these overshadowed the relationship between traffic accidents and their causes, roads, and vehicles (physical factors), which is explained by the significant deterioration in the traffic regulations and laws.

Keywords: Traffic accidents · Human and physical factors · Multiple linear regression model · Iraq

1 Introduction

The system of movement is a complex operation; any defect in a part of it leads to severe consequences and accidents. Traffic accidents have adversely affected most of the economies of developed and developing countries, which is considered one of the most significant factors for the loss of human and material resources. Developed countries have taken remarkable steps to reduce this problem, while the problem has been exacerbated and increased in developing countries.

1.1 Traffic Accidents

Traffic accidents are one of the major problems associated with the modern era and with the widespread use of transport modes. These incidents have many negative economic and social impacts, and these impact increasing with the rise in the number of

© Springer Nature Switzerland AG 2020
E. Macioszek and G. Sierpiński (Eds.): Modern Traffic Engineering in the System Approach to the Development of Traffic Networks, AISC 1083, pp. 190–201, 2020.
https://doi.org/10.1007/978-3-030-34069-8_15

vehicles and their speed. The number of such incidents was minor with the first incident which killed two people in 1896 [1]. While now according to the latest estimates of scientific studies, it kills more than 1.2 million people a year and injures about 50 million others, which cost about 3.6% of the global mortality, depending on WHO [2], and it is expected to double by 2030. The share of developing countries in these incidents is 86%. Around 10% of hospital beds are occupied by traffic victims; therefore, it is competing with other health and security causes of death worldwide [3, 4]. The road's fatality incident rate in some developing country like Kenya during 2009, was of 59.96 per 100,000 inhabitants. It is high when compared to the global rate of 21.5 and 10.3 per 100,000 inhabitants for low and high-income countries, respectively [5]. In Ghana, an average of four souls are killed daily by road crashes in 2005, which is 16% more than 2004 statistics [6]. In Bangladesh, the rate of annual fatality in roads accident was 85.6 per 10,000 vehicles in comparison to 3 per 10,000 vehicles in developed countries [7]. Human losses are expected to increase in the Middle East by the year 2020 by 7% of the total dead in the world [3]. Another study by the WHO indicated that it is expected that the rate of traffic accidents in developed countries for the year 2030 will decrease by 28% [4].

1.2   Cost of Traffic Accidents

Traffic accidents involve losses of up to $500 billion per year. Which equivalent to about 2% of global output [4]. In Ghana alone, which is considered a low-income country, National Road Safety Commission's statistics shows that there is a loss of 1.7% of the national product, which is over $230 million per year [6]. Low-income countries incur estimated annual losses of $65 billion to $100 billion. These losses include both material and costs losses and efforts of care for those injured by these accidents [7]. There is a lack of documentation of traffic accidents and statistics in developing countries. Data should be viewed with caution because their quality is affected by not reporting all incidents. As well as the incomplete retrieval of reported incidents from police files [8]. Society cost such as loss of economically active people who may have contributed to various economic activities, loss of youth and educated people, loss of government and families resources, insurance companies losses, damage to property are inestimable. Valuing things like psychosocial influence on victims is a difficult task. Suffering, loss of life, and injuries linked with road traffic accidents are hard to assign a monetary value [8]. As for the traffic accidents in the Arab countries, the statistics of the Arab League indicated that about 26 thousand dead and more than a quarter of a million injured annually. In addition to economic losses amounting of $60 billion annually [4].

1.3   Causes of Accidents

Traffic accidents include three main types: 1 - Rollover: which is caused by an imbalance of vehicle and rolling. 2 - Run over: which is caused by the collision of the cars with the pedestrians. 3 - Collisions: which classify into, a collision of vehicles between each other, and collision of vehicles with road barriers. The traffic accident between more than one road user can be attributed as showing in Fig. 1.

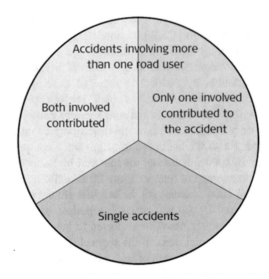

Fig. 1. The ratio of responsibility of road user in the traffic accident (Source: [9])

The main factors causing accidents can be classified into two categories: human and physical factors.

1.3.1   Human Factors

Kenya Roads Board's report in 2009 indicates that human factors caused of 85.5% of all road accidents. The share of motorists and motorcyclists was (43.6%), pedestrians (24.8%), riders (4.8%) and cyclist (10.3%) [5]. The driver alone is directly responsible for the vehicle's driving control. The age, culture, driving experience, behavior, and mental and physical health are indicators of accidents [10]. Neglect and lack of know-how to use safety equipment constitute a significant indicator of the severity of the accident. The lack of attention caused by drowsiness has caused many accidents, as not to sleep for 18 h and then driving the car is equivalent to drinking alcohol by 0.05% in the blood, which is more than allowed by law [3].

Factors such as traffic culture, discipline, and compliance with traffic guidelines affect the proportion of pedestrians involved in traffic accidents [10].

1.3.2   Physical Factors

The most common physical factors that caused accidents are vehicles and roads, as well as some natural obstacles such as rocks or trees that appear abrupt on the streets. The road represents an essential component of road traffic accidents through its design, surface, planning characteristics, and maintenance. The intersections, detours, slopes, pedestrian walkways, traffic control, and others, all are factors affecting the proportion of traffic accidents [3, 5].

The vehicle is the tool that is causing traffic accidents, either through poor driving or fault in one of its parts. Improvements in the vehicle production sector have helped to reduce the risk of using these vehicles, DTU Transport in 2012, indicate that increases in safety instruments have decreased the number of injuries. Which confirmed by the growing use of airbags and anti-skid technology (ESC), and so on, which are among the techniques associated with minimizing the risk of personal injury [11].

1.3.3   Other Factors Affect Traffic Accidents

Traffic accidents are affected by many different factors such as weather conditions, day and night light, traffic volume and composition, traffic laws and regulations (traffic management) as well as planning style of the city. Sir Alker, in 1942, raised the problem of traffic accidents based on a planning problem that reflected poor planning. He described the city, which plans to leave its inhabitants to be killed and injured in large numbers as an unplanned city [11]. However, it is difficult to distinguish between the causes of accidents, one or more factors may be involved in causing the accident, as illustrated by the Fig. 2.

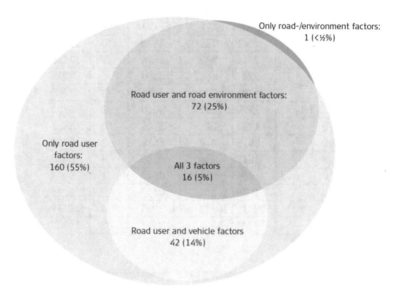

Fig. 2. How often factors (road user, vehicle, road, and environment) are found in individual accidents (Source: [9])

2   Case Study and Data Source

In this paper, two time-series of Iraqi traffic accidents were studied. The first was characterized by being more organized. A comparison was made between these two periods in order to identify the most important reasons related to the type of incident and the cause of the occurrence. The available Iraqi statistical publications for the years 1979–2000, and 2010–2017 were adopted as a source of data, as shown in Tables 1, 2, 3 and 4.

Table 1. Traffic accidents recorded according to the nature of the accident in Iraq for the period 1979–2000 (Source: [12]).

Year/Accident	Crash	Rollover	Run over	Others	Sum
1979	12366	2202	7291	141	22000
1980	16998	2302	9557	128	28985
1981	16405	2498	10273	109	29285
1982	16806	2581	9140	159	28686
1983	17274	3144	11417	153	31988
1984	16815	3008	13126	127	33076
1985	14994	2907	13960	142	32003
1986	15787	2946	15596	84	34413
1987	12025	2655	14105	101	28886
1988	8184	1964	11460	61	21669
1989	10785	2230	11921	54	24990
1990	10905	2171	12177	57	25310
1991	4241	741	6520	10	11512
1992	4814	739	6816	12	12381
1993	2227	466	6299	6	8998
1994	1797	322	3644	2	5765
1995	1379	232	3610	2	5223
1996	1338	249	3744	0	5331
1997	1913	219	4618	2	6752
1998	1811	236	4349	4	6400
1999	1964	275	4543	1	6783
2000	1989	358	4439	9	6795
Sum	192817	34445	188605	1364	417231
Average	8764	1565	8572	62	18965

Table 2. Traffic accidents recorded according to the nature of the accident in Iraq for the period 2010–2017 (Source: [13]).

Year/Accident	Crash	Rollover	Run over	Others	Sum
2010	4102	1011	3661	87	8861
2011	4771	1161	4025	125	10082
2012	5133	1320	4174	84	10711
2013	4568	1288	3793	76	9725
2014	4288	993	3442	91	8814
2015	4213	1000	3405	218	8836
2016	4242	946	3431	144	8763
2017	4446	918	3267	193	8824
Sum	35763	8637	29198	1018	74616
Average	4470	1079	3649	127	9327

Table 3. Traffic accidents recorded according to the cause of the accident in Iraq for the period 1979–2000 (Source: [12]).

Year/Details	Road	Car	Driver	Pedestrian
1979	765	2375	18074	786
1980	958	2654	24377	996
1981	766	2293	25079	1147
1982	863	2396	24470	957
1983	896	2653	27309	1130
1984	696	2378	28535	1467
1985	877	2414	27334	1378
1986	1059	2971	28928	1455
1987	1238	1999	24189	1460
1988	605	1756	18064	1244
1989	748	1898	20749	1595
1990	386	1769	21169	1986
1991	108	610	9798	996
1992	116	629	10718	918
1993	92	642	7567	697
1994	22	284	5046	413
1995	24	223	4551	425
1996	50	290	4597	394
1997	28	318	6079	327
1998	45	294	5692	369
1999	90	374	5863	456
2000	132	383	5809	471
Sum	10564	31603	353997	21067
Average	480	1436	16090	957

Table 4. Traffic accidents recorded according to the cause of the accident in Iraq for the period 2010–2017 (Source: [13]).

Year/Details	Road	Car	Driver	Pedestrian
2010	707	1491	6143	280
2011	851	1638	6956	348
2012	911	1816	7277	427
2013	910	1568	6681	325
2014	856	1136	6446	220
2015	841	1101	6496	205
2016	647	1064	6773	140
2017	682	1251	6351	254
Sum	6405	11065	53123	2199
Average	800	1383	6640	274

2.1   Traffic Accident Indicators

From Tables 1 and 3, it can be observed that the collision was most of the traffic accidents for the period 1979–2000, with an annual rate of 8764.41 and 46.2%. The incidence of run-over was 45.2%, and rollover was 8.2%. While the percentages for the statistics in the period 2010–2017, 47.9% for collisions, 39.1% for the run-over, 11.6% for the rollover. From Tables 2 and 4, which related to the causes of accidents, human factors were the predominant cause of accidents. The accidents caused by the driver for the period 1979–2000 was 84.8%, caused by the pedestrian 5% and 7.6% -2.5% due to vehicle and road, respectively. While the statistics for the period 2010–2017, showed that 75% of the incidents were due to human factors.

3   Method of Analysis

The multiple linear regression model was used to determine the relationship between the nature of traffic accidents and the human and physical factors causing them. In order to find the causal relations between the variables of the study, and to clarify the relationship between the variable that represents the traffic accidents (Y_i) and the independent variables that represent the factors causing the traffic accidents (X_i) in the model. Depending on the following decision-making base:

- null Hypothesis H0: $\beta i = 0$ (no effect of the independent variable on the approved variable),
- alternative Hypothecs H1: $\beta i \neq 0$ (there is an effect of the independent variable on the approved variable).

Statistical methods were chosen according to the nature of the research hypotheses, using the Statistical Package for Social Sciences (SPSS) to analyzing the data, which based on the statistical significance level $(0.05 \leq a)$. Where: (a) represents the probability value of the permissible error level, that is, the degree of confidence in the model as a probability value (95%).

4   Results

The study times were divided into two periods (1979–2000) and (2010–2017), the variables studied were classified as follows: ($X1$ = Road, $X2$ = Car, $X3$ = Driver, $X4$ = Pedestrian, and $Y1$ = Crash, $Y2$ = Roll over, $Y3$ = Run over, $Y4$ = Other). The analysis of the data in the Tables (1, 2, 3 and 4) produced the results indicated in the Tables 5, 6, 7, 8, 9, 10 and subsequent discussions.

4.1   First Period

By applying the regression analysis model to the first-period data (1979–2000), a set of indicators was found. Table 5. The Corrected selection factor for all Estimated Models was ($R^2 \sim 0.98$), that means the explanatory variables explain the (98%) of the variance, which shows the contribution of the independent variables Xi in the interpretation of the change in the variable Y_i (traffic accidents).

Table 5. Estimation equations and the values of the factors and constants resulting from the application of the regression model for the first period 1979–2000.

Dependent variables	Independent variables (X1, X2, X3, X4) and the constant parameter (β0)					R	R^2	Value F	Sig.	H0
Y1	β0	B1	B2	B3	β4					
Value of parameter	−1047.17	−2.73	2.25	0.67	−3.07	99%	97%	233.8	0.00	Reject
Value T	−2.05	−1.74	1.98	5.1	−3.53					
Sig.	0.5	0.1	0.6	0.0	0.03					
The estimated equation: $\hat{Y}1 = -1047.17 - 2.73X1 + 2.253X2 + 0.67X3 - 3.07X4$										
Y2	β0	B1	B2	B3	β4	99%	98%	311.12	0.00	Reject
Value of parameter	−306.68	0.27	0.22	0.08	0.12					
Value T	−3.89	1.12	1.28	4.04	0.92					
Sig.	0.01	0.2	0.2	0.0	0.4					
The estimated equation: $\hat{Y}2 = -306.68 - 0.27X1 + 0.22X2 + 0.08X3 + 0.12X4$										
Y3	β0	B1	B2	B3	β4	97%	94%	85.87	0.00	Reject
Value of parameter	1414.16	3.36	−1.34	0.23	4.04					
Value T	2.78	2.15	−1.19	1.77	4.66					
Sig.	0.1	0.0	0.3	0.1	0.0					
The estimated equation: $\hat{Y}3 = 1414.1 + 3.36X1 - 1.34X2 + 0.23X3 + 4.04X4$										
Y4	β0	B1	B2	B3	β4	95%	88%	37.1	0.00	Reject
Value of parameter	−7.12	0.01	0.02	0.01	−0.06					
Value T	−0.61	0.16	0.97	1.86	−2.98					
Sig.	0.5	0.8	0.3	0.1	0.0					
The estimated equation: $\hat{Y}4 = -7.12 - 0.01X1 + 0.02X2 + 0.01X3 - 0.06X4$										
Av. Y	β0	B1	B2	B3	β4	100%	100%	281650.0	0.00	Reject
Value of parameter	52781	0.91	1.15	0.99	1.04					
Value T	2053	11.53	20.21	149.56	23.65					
Sig.	0.0	0.01	0.00	0.1	0.1					
The estimated equation: $\hat{Y} = 52781 + 0.91X1 + 1.15X2 + 0.99X3 + 1.04X4$										

Table 6. The descending sequence of impact values.

Range	Independent variables (X)	Effect of (Av. Y)
1.	Car	B2 = 1.15
2.	Pedestrian	B4 = 1.04
3.	Driver	B3 = 0.99
4.	Road	B1 = 0.91

Table 7. The descending sequence of the significance of the variables.

Range	Independent variables (X)	Effect of (Av. Y)
1.	Driver	149.56
2.	Pedestrian	23.65
3.	Car	20.21
4.	Road	11.53

Table 8. Estimation Equations and the values of the factors and constants resulting from the application of the regression model for the second period 2010–2017.

Dependent variables	Independent variables ($X1$, $X2$, $X3$, $X4$) and the constant parameter ($\beta0$)					R	R^2	F value	Sig.	$H0$
Y1	$\beta0$	B1	B2	B3	B4					
Value of parameter	304.92	0.43	−0.24	0.55	3.81	99%	96%	41.33	0.00	Reject
Value T	0.43	0.84	−0.56	6.15	2.09					
Sig.	0.6	0.5	0.6	0.0	0.1					
The estimated equation: $\hat{Y}1 = 304.92 + 0.43X1 − 0.24X2 + 0.55X3 + 3.81X4$										
Y2	$\beta0$	B1	B2	B3	B4	98%	91%	18.94	0.01	Reject
Value of parameter	−1580.40	1.27	1.26	0.12	−3.83					
Value T	−3.32	3.71	4.43	1.1	−3.15					
Sig.	0.00	0.0	0.0	0.1	0.1					
The estimated equation: $\hat{Y}2 = −1580.40 + 1.27X1 + 1.26X2 + 0.12X3 − 3.83X4$										
Y3	$\beta0$	B1	B2	B3	B4	97%	88%	13.26	0.03	Reject
Value of parameter	−276.40	0.42	1.65	0.31	−3.03					
Value T	−0.24	0.49	2.36	2.12	−1.01					
Sig.	0.8	0.6	0.0	0.1	0.4					
The estimated equation: $\hat{Y}3 = −276.40 + 0.42X1 + 1.65X2 + 0.31X3 − 3.03X4$										
Y4	$\beta0$	B1	B2	B3	B4	70%	45%	0.74	0.6	Accept
Value of parameter	469.07	−0.16	−0.42	0.02	1.24					
Value T	0.78	−0.36	−1.18	0.22	0.82					
Sig.	0.4	0.7	0.3	0.8	0.4					
The estimated equation: $\hat{Y}4 = 469.07 − 0.16X1 − 0.42X2 + 0.02X3 + 1.24X4$										
Av. (Y)	$\beta0$	B1	B2	B3	B4	91%	61%	3.71	0.01	Reject
Value of parameter	16833.27	−40.47	−4.07	0.85	95.28					
Value T	0.51	−1.71	−0.21	0.21	1.13					
Sig.	0.6	0.5	0.9	0.8	0.3					
The estimated equation: Av.(Y) = 16833.27 − 40.47X1 − 4.07X2 + 0.85X3 + 95.28X4										

Table 9. The descending sequence of impact values.

Range	Independent variables (X)	Effect of (Av. Y)
1.	Pedestrian	B4 = 95.28
2.	Driver	B3 = 0.85
3.	Car	B2 = −4.07
4.	Road	B1 = −40.47

Table 10. The descending sequence of the significance of the variables.

Range	Independent variables (X)	Effect of (Av. Y)
1.	Pedestrian	1.13
2.	Driver	0.21
3.	Car	−0.21
4.	Road	−1.71

The value of the F test for all estimated equations is significant as well as for the (Av = 281650.0) and this reflects the strength of the interpretation of variables in the Estimated Model.

The most significant effect of the Av. model (Av. Y) was the car, (Sig. = 0.00) which may be since it is the era of the Iran-Iraq War and the period of economic blockade. It was challenging to obtain the necessary spare parts for the maintenance of vehicles, especially tires, which may have contributed to the increase in accidents caused by the car as demonstrated by the model. The average impact of cars ($\beta 2 = 1.15$) was a relative increase of 15% compared with the average. That means that an increase of one unit of vehicles will lead to a rise in the accident (1.15%) above average. While the least impact on traffic accidents was the road with an impact rate of ($\beta 1 = 0.91$) less than the average (9%). Table 6 shows the descending sequence of the values of the effect of independent variables on the adopted variable.

The driver was one of the most influential factors in the percentage of traffic accidents, among other variables, when analyzing the level of significance of the variables using a t-test. Where (t) was (149.56) which means the accidents due to the driver have a significant impact on the rate of traffic accidents when tested separately from other variables, which indicates the lack of efficiency and experience of drivers who caused the accidents. While the least significant variable in term of influence was the road, which effects by 11.53, this indicates that the roads were in good condition. Table 7 shows the descending sequence of the significance of the variables.

4.2   Second Period

In the second period (2010–2017), after applying the regression model to its data, a sort of variance and difference from the previous study period was found. Table 8. The Corrected selection factor for all Estimated Models was ($R^2 \sim 0.76$), that means the explanatory variables explain the (76%) of the variance, which shows the contribution of the independent variables Xi in the interpretation of the change in the variable Yi (traffic accidents).

The value of the (F - test) for all estimated equations is significant as well as for the (Av = 3.71), and this reflects the strength of the interpretation of variables in the Estimated Model

The most significant effect of the Av. model (Av. Y) was the pedestrian (Sig. = 0.3), which is a sign of the considerable deterioration in the traffic system and compliance with the traffic laws that characterized this period whether by pedestrians or drivers. The streets lack the essential elements of traffic, such as traffic signs, pedestrian crossing lines, and others because of the war and destruction that have affected Iraqi cities. The average impact of pedestrians in traffic accidents was ($\beta 4$ = 95.28).

This means that an increase of one unit of the walker will lead to a rise in the accident (95.28%). While the least impact on traffic accidents was the road with an impact rate of ($\beta 1$ = −40.47). That means that increasing ways per unit reduces pedestrian density by (40.47). Despite the unfortunate situation that Iraq's roads have become at this time. However, traffic accidents were likely attributed to more critical reasons such as speeding and non-adherence to traffic regulations, which reduced the impact of roads on traffic accidents. Table 9 shows the descending sequence of the values of the effect of independent variables on the adopted variable.

The pedestrian was the most influential factors in the percentage of traffic accidents, among other variables, when analyzing the level of significance of the variables using a t-test. Where (t) was (1.13), which means the accidents due to the Pedestrian have a significant impact on the rate of traffic accidents when tested separately from other variables. While the least significant variable in term of influence was the road, which effects by 1.71, this supports the results of the regression model. Table 10 shows the descending sequence of the significance of the variables.

5 Conclusions

In this study, two periods studied: the first was from 1979–2000, the second was between 2010–2017. Traffic conditions varied for each of these periods, which in turn led to a difference in the relation between the various traffic accidents, and the causes of these incidents of human (driver and pedestrian) and physical (vehicle and road) factors.

According to the results of the regression analysis of the studied time series, the effect of the vehicle was significant according to the general model data in the first period. The period between 1979–2000 was characterized by technical difficulties related to the maintenance and rehabilitation of vehicles, as a result of the outbreak of the first and second Gulf wars and the subsequent economic blockade on Iraq. In that period, most of the vehicle spare parts were manufactured locally with low quality, which made the car a prominent factor in traffic accidents.

The second period was accompanied by extensive military operations in the country, characterized by the absence of traffic regulation and non-compliance with the application of traffic laws. Where most of the traffic light stopped working, and many of the traffic signs, roadway marking, and pedestrian crossing lines disappeared from the roads. All of which led to a significant overlap in movement of pedestrian and vehicles, which led to many traffic accidents that appeared clearly in the results of the model.

References

1. Hussein, R., Bilal, B.: Behavioral disorders of young drivers and their relationship to traffic accidents. Master thesis, Aljilali University, Algeria (2018)
2. World Health Organization Statistic, Geneva. https://www.who.int/gho/publications/world_health_statistics/en/
3. Rashed, W.Q.: Economic costs and losses of traffic accidents in Basrah City for the years 2004–2009. Al Ghari J. Econ. Adm. Sci. **14**(3), 972–997 (2017)
4. Dahd, S.: Traffic accidents in Diqar Governorate, causes and solutions. J. Fac. Basic Educ. Educ. Hum. Sci. **20**, 639–655 (2015)
5. Daniel, O.: Exploring the major causes of road traffic accidents in Nairobi County. Master thesis. College of biological and physical sciences-school of mathematics, University of Nairobi, Nairobi (2016)
6. Haadi, A.R.: Identification of factors that cause severity of road accidents in Ghana: a case study of the northern region. Int. J. Appl. Sci. Technol. **4**(3), 242–249 (2014)
7. Boni, A.J., Chowdhury, T.R., Das, S.S.: A study on causes of road accidents at Dhaka to Comilla highway. Asian J. Innov. Res. Sci. Eng. Technol. **1**(9), 6–13 (2016)
8. Osoro, A.A., Ng'ang'a, Z., Yitambe, A.: An analysis of the incidence and causes of road traffic accident in Kisii, Central District, Kenya. J. Pharm. **5**(9), 41–49 (2015)
9. Danish Road Traffic Accident Investigation Board, why do road traffic accidents occur? http://www.hvu.dk/SiteCollectionDocuments/HVUdec14_UK_HvorforSkerUlykkerne.pdf
10. Nusrat, Y., Bahaa, A.A.: Characteristics of Traffic Accidents in Cities. Study No. 591. Iraq Ministry of Planning Study, Iraq (1988)
11. Al-Wasti, A.: Quantitative analysis of road accidents in the Iraqi City. Master thesis. Urban and Regional Planning Center. University of Baghdad, Bagdad (1987)
12. Directorate of Transport and Communications: Traffic Accidents Report. Central Statistical Organization, unpublished data, excel sheet (2000)
13. Directorate of Transport and Communications: Traffic Accidents Report. Central Statistical Organization, unpublished data, excel sheet (2017)

A Case Study of High Power Fire and Evacuation Process in an Urban Road Tunnel

Aleksander Król[1(✉)] and Małgorzata Król[2]

[1] Faculty of Transport, Silesian University of Technology, Katowice, Poland
aleksander.krol@polsl.pl
[2] Faculty of Energy and Environmental Engineering,
Silesian University of Technology, Gliwice, Poland
malgorzata.krol@polsl.pl

Abstract. Urban road tunnels become recently inherent components of road network. They facilitate urban transport and allow for arrangement of urban space. However, hereby new dangers appeared and the most serious one is the fire outbreak in an urban tunnel with congested traffic. Since a tunnel fire commonly develops quickly and in an unpredictable way the threatened people have to undertake appropriate actions to save themselves. All tunnel systems must support the self-rescue process, but in a case of high power fire the time factor is of crucial importance. The paper presents a few scenarios of tunnel fire and subsequent evacuation in a real urban tunnel. The fire development was simulated using Computational Fluid Mechanics (CFD) methods and the evacuation process was modeled by the PATHFINDER software.

Keywords: Road tunnel · High power tunnel fire · Evacuation · Fire scenarios

1 Introduction

The fire growth in a road tunnel is always a huge threat. The large amounts of toxic smoke occur and make the evacuation of people very difficult. The fire outbreak and even a not fully developed fire causes a significant part of the tunnel filled with toxic and hot fire gases. This dangerous for a human health and life zone is expanding fast, covering the successive parts of a tunnel in a few minutes. Therefore, it is extremely important that people in the tunnel undertake the self-evacuation in the initial phase of a fire development. For this to happen all the tunnel safety systems have to operate properly and support the self-rescue of the people [1–4].

Many factors affect the safety of people during the self-evacuation from a tunnel. Apart some human aspects like panic, anxiety, speed of decision-making or general fitness there are lots of factors connected with safety systems of a tunnel. Most of all fire detection, fire alarm and ventilation systems should be mentioned here. The tunnel type (bi or unidirectional), the traffic intensity and the fire location can also influence the evacuation process. A fire detection system should be able to distinguish the fire symptoms from other phenomena, which can be the natural effects of common tunnel

E. Macioszek and G. Sierpiński (Eds.): Modern Traffic Engineering in the System Approach to the Development of Traffic Networks, AISC 1083, pp. 202–215, 2020.
https://doi.org/10.1007/978-3-030-34069-8_16

operation. Tests of different fire detection methods showed that response time of fire detection systems is rather short, it did not exceeded 60 s [5]. The researches in a real tunnel confirmed it [6]. On the other hand, there are mentioned in the literature cases, when this time was significantly longer and was of order of a few minutes [5]. An especially difficult challenge for fire detection systems are fires which undergo an incubation period developing in hiding, in such circumstances it could take even up to 8 min [7].

Following the activation of the fire detection system the ventilation system is switched on in a fire mode. The regulations in the European countries concerning the applied ventilation system don't determine unambiguously when a mechanical system is required. According to the German guidelines RABT the natural ventilation systems except a few cases are allowed for tunnels not longer than 400 m [8]. Therefore, longer tunnels are usually equipped with mechanical ventilation. It could be implemented as longitudinal or transverse system or a combination of both. The selection a relevant system depends on a tunnel length, traffic mode (bi or unidirectional) and traffic intensity [3, 9]. The general principle of the operation of a ventilation system is to keep as large as possible part of a tunnel free of smoke for as long time as possible.

A fire is developing very quickly in a road tunnel. There are even possible fire jumps to a relatively distant vehicle [10]. It is very difficult to predict the intensity of a tunnel fire. According to the real tests the power of a fire is from 3 MW to 30 MW or even 70 MW for a passenger car or heavy goods vehicle (HGV) respectively. Some works reported even larger values.

The number of people whose health and life could be threatened in a case of fire in a road tunnel depends on the number of vehicles, which are trapped inside and the filling of vehicles. In turn the number of trapped vehicles depends on the traffic conditions, the traffic mode, the number of lanes and the location of a road accident. Therefore determination of the number of trapped vehicles is a complex task. To solve it some theoretical models or numerical simulation of road traffic can be applied, additionally statistical data based on traffic measurements must be provided [11, 12]. Often, however, it happens that in this type of analysis the traffic congestion in the tunnel is created manually based on theoretical knowledge.

As was earlier mentioned the fire development in a road tunnel is a very quick and unpredictable process. The chances of surviving are determined by decisions and actions undertaken by the threatened people themselves in as short time as possible. The time needed for all threatened persons to leave the dangerous zone has to be smaller than the time of worsening the conditions below the critical values.

Since the theoretical description of the evacuation process has to take into account the human behavior it is a difficult problem. People being in stressing circumstances may not perceive some symptoms of a danger or improperly interpret them [13]. The main challenge for an alarm system is to urge the endangered people to leave their vehicles and move toward safe places [14–16]. The message should be clear and must show no doubts on the threat [17]. A problem was observed that people not always chose the shortest escape route [18], thus the appropriate marking of escape routes is of crucial significance [19]. As the evacuees often follow by the behavior of others, it leads to forming groups of persons behaving similarly [20, 21].

There are many approaches to model the evacuation process. The simplest one is just to estimate selected evacuation parameters. Macroscopic models of evacuees' movement take into account the geometry of escape routes in a way similar to modeling the fluid flow, whereas microscopic behavioral models consider behaviors of particular persons [22, 23].

All approaches of the evacuation modeling suffer to the shortage of real data [24]. Therefore many assumptions on the human behavior are not sufficiently validated; it could decrease the reliability of obtained results [25–27]. However, some available software packages for evacuation simulation are recognized as trustworthy. The realistic reproducing of the evacuees' behavior requires accounting for many individual parameters: movement speed, shoulder width, overall fitness, sex, age and possible delays; these parameters are often introduced as random variables with given distributions [28–30].

The plethora of available software leads to the question whether the obtained results are mutually coherent. This issue was examined by Ronchi [31, 32], who compared the most often applied programs: Pyrosim + EVAC, STEPS and PATHFINDER. He found all the examined programs to give the similar overall image of the evacuation process. Since STEPS and the used version of PATHFINDER were not coupled with the fire development model, it was a need to create manually additional zones of dangerous conditions. It was confirmed by Nan, who compared the real data from an evacuation experiment with simulation results of Pyrosim + EVAC and PATHFINDER. Although, the participants of this experiment did not know its scenario, they were aware of its aim. The results provided by both software packages generally agreed with the experiment [33], but PATHFINDER was the more accurate one if only the additional data on fire condition were introduced into the model.

In the presented work a few high power fire and subsequent evacuation scenarios in an urban road tunnel with congested traffic are analyzed. Since the examined tunnel models an actual facility, at the beginning the article presents a short description of the tunnel, its localization and equipment. Using numerical modeling with Fire Dynamics Simulator (FDS), the development of a fire with two different powers was analyzed; two different fire detection times were also checked in the calculations. The PATHFINDER program was then used to analyze the evacuation process. The current version of PATHFINDER, which was applied, was able to couple itself with the data on fire development provided by FDS.

2 Tunnel Description

The Katowicki Tunnel is a segment of national road no 79 and is located just in downtown of Katowice. It goes almost latitudinally (Fig. 1). It contains two tubes with unidirectional traffic. The length of northern tube is 657 m, the southern one is a bit shorter and its length is 650 m. The cross-section shape is rectangular, the height of both tubes is 6.4 m and width is 12 m. There are 3 lanes in each tube. The ventilation system consists of 7 pairs of axial fans in each tube. The tunnel is equipped with 4 evacuation exits, placed every 150 m. According to the General Traffic Measurement 2015 [34] the daily traffic volume is about 100 000 vehicles. Heavy congestions appear in the tunnel often at rush hours or in a case of a road accident. In such a situation all lanes at entire tunnel length are filled with vehicles.

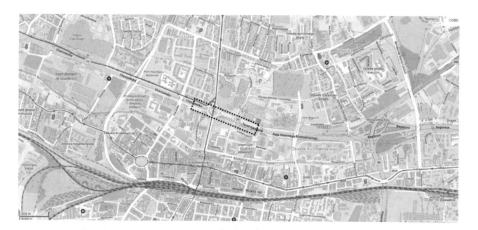

Fig. 1. Tunnel location

3   Modeling of the Fire Development in a Road Tunnel

The work was focused on high power tunnel fires, because such a fire is a danger to all people trapped in a tunnel. If a single passenger car is burning the expected fire power is of order of 4–6 MW [16, 35] and the tunnel ventilation system should rather deal with the effects of such low power fire [36]. The situation appears quite different when considering a fire of a bus or a HGV. A fire of a typical bus results in the heat release rate (HRR) of 20–30 MW [27]. The fire power of HGV strongly depends on its load: it can be below 20 MW for inflammable goods, about 70–100 MW for combustible materials like wood or margarine and even over 300 MW for fuel tanks [37].

The main threat to the people in the tunnel is smoke, which contains toxic components and substantially limits the visibility [27]. In the first phase of the tunnel fire the smoke goes up reaching the ceiling. Then it spreads in both directions, however in the presence of natural or forced air flow in the tunnel its extent on windward is smaller than on leeward. The heat is transferred to the tunnel walls and the fresh air is sucked into the smoke plume, therefore the smoke is cooled, the buoyancy forces are weakening and the smoke descends. In such a way, after a short time period (5 to 10 min) the whole tunnel space becomes filled by smoke [38, 39]. This is shown in Fig. 2; since each fire develops in a different way, the provided time stamps should be regarded as indicatory.

The fire development and the smoke spreading could be modeled using CFD methods. In this work FDS software was used. The computational domain corresponded to the tunnel interior and was divided into a number of rectangular cells (816000). Both portals were opened surfaces with pressure boundary condition. A low pressure difference of 1 Pa was applied between portals to simulate natural air flow due to the influence of the external wind, which is a common phenomenon in tunnels.

The heat flux of the surface fire source was the same in all cases (750 kW/m^2). It is due to obtain actual temperature distribution above the fire source [40].

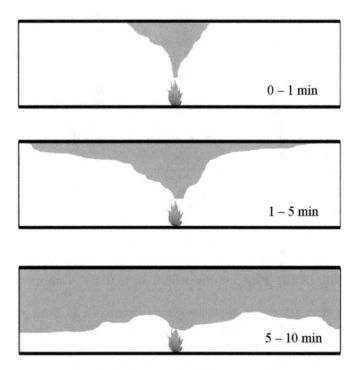

Fig. 2. Smoke spreading during a tunnel fire

4   Examined Scenarios of Fire Accident and Evacuation

Two fire cases were checked: a typical bus and a HGV with combustible load. Different fire powers were obtained by changing the area of fire source, 24 m^2 for HRR = 18 MW and 96 m^2 for HRR = 72 MW respectively. According to the PIARC report [41] a linear fire growth was assumed and fire growth times were adopted (120 s for the lower HRR and 240 s for the higher HRR). Although some papers report longer time periods, even up to 15 min [42], the adopted values are in compliance with commonly recognized data [5]. In all scenarios the traffic accident and resulting fire was located just in the middle of the tunnel - at a distance of 330 m from both portals in the middle lane.

The examined scenarios assumed that in given time period after ignition the fire was detected, then the alarm was announced, tunnel entrance closed and the ventilation system switched into the fire mode. Two detection delays were checked: delay of 60 s according to the RABT standards [8] and delay of 120 s also admissible by design standards [28].

When the ventilation system is in the fire mode, the axial fans start to operate. The fans capacity is 25 m³/s each; it is achieved in 30 s after switching on. Since the main aim of ventilation system operation is to keep as large volume of tunnel interior free of smoke for as long time as possible, for each tunnel so called critical velocity is determined [9]. It is a value of velocity of forced air flow, which should prevent the back flow of hot fire gases. The examined scenarios of fire development are shown in Table 1.

Table 1. Examined scenarios of fire development

Case	Fire power [MW]	Detection delay [s]
A	18	60
B	72	60
C	18	120
D	72	120

For all examined cases the same model of traffic congestion in the tunnel was assumed. All the vehicles being inside the tunnel are regarded as trapped. The sketch of the formed congestion is shown in Fig. 3.

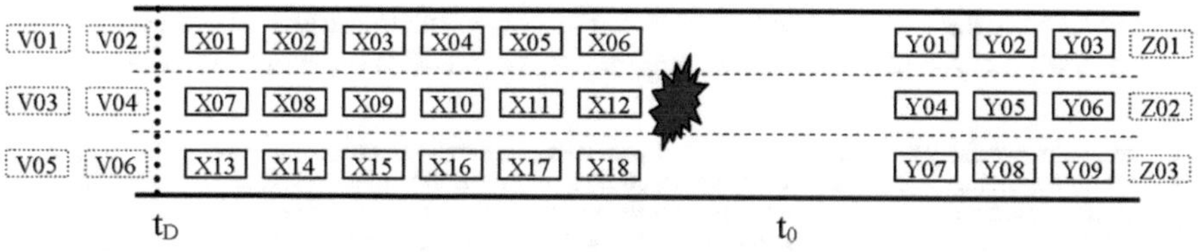

Fig. 3. Traffic congestion in a tunnel (t_0 denotes the moment of accident, t_D denotes the moment of fire detection and tunnel closing)

There are the vehicles which got stuck because of the accident (X** group). The following vehicles (V** group) are not able to enter the tunnel due to congestion and tunnel closing. Some of vehicles preceding the accident manage to leave the tunnel despite the congestion (Z** group). Passengers of these both groups are not taken into account. The last group contains the vehicles, which are not able to leave due to congestion (Y** group). The free space between the accident location and the vehicles of Y** group was estimated assuming a traffic speed of 10 km/h at the congestion state.

The average number of passenger per vehicle was adopted according to Mikame: 1.4 per a passenger car, 1.3 per a HGV and 50 per a bus [43]. Next, taking into account the generic structure of the traffic in the Katowicki Tunnel [34] the traffic congestion was randomly constructed. It contained 193 vehicles, including 160 passenger cars, 30 HGVs and 3 buses. The total number of trapped people was 398.

A passenger in a trapped car has to make the decision on evacuation by himself. There are three groups of evacuees [27]:

- one can see the accident or the fire outbreak. It concerns the people in cars close to the accident. The average time needed to start to escape is about 45 s with standard deviation of 15 s,
- the majority of people wait for alarm announcement; then they have to recognize the situation and make a decision. This group starts to move after average 30 s with the same standard deviation,
- some people start to escape seeing the evacuees from the first group. This was simulated by setting a minimum time constraint for second group to a value lower than the alarm moment.

It was assumed that the velocity of evacuating people was a random variable of normal distribution. The mean value was 1.2 m/s and the standard deviation was 0.2 m/s.

5   Results

The critical velocity for Katowicki Tunnel was calculated as 3.2 m/s [9]. Despite such intense airflow was provided in all examined cases, the back flow of smoke appeared. The higher the fire power is the stronger the back flow (in the opposite direction to the fans operation) is. The range of the back flow is also larger for the longer delay of fire detection. It is because the fans started later to operate. The spread of smoke at 150 s of simulation is shown in Fig. 4. Additionally the emergency exits (E1–E4) are marked in the Fig. 4.

Fig. 4. The spread of smoke at 150 s of simulation for all examined cases

This phenomenon is of major importance for the people trapped in vehicles before fire source (X** group). The operation of fans and natural flow are forcing the smoke to move towards the people of Y** group, but they have a bit longer time to recognize the situation.

The PATHFINDER software creates individual files for each evacuating person. Such file contains detailed history of a person, which includes his position and surrounding conditions like temperature, concentration of carbon monoxide and visibility. It was assumed that a person who encountered the smoke density resulting in visibility less than 10 m was in a serious danger. The evacuation process for scenarios A–D with such defined safety criterion was visualized in Figs. 5, 6, 7 and 8. There are individual paths of evacuees visible as grey lines. The abscissa of each point determines the location of a person along the tunnel length relatively to the fire source location; the ordinate determines the time moment. Black bold indicates the threat of the low visibility (less than 10 m).

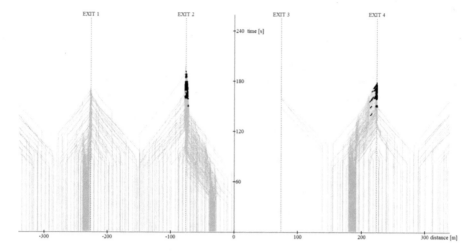

Fig. 5. Paths of the evacuation for case A

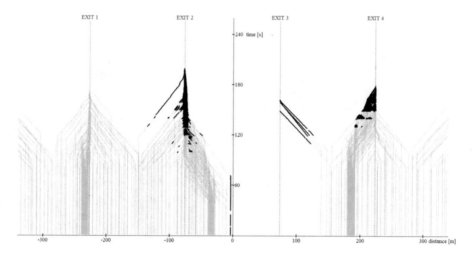

Fig. 6. Paths of the evacuation for case B

The vertical segments of paths indicate the state of immobility of a person. Such state can be observed at the beginning of evacuation - before the decision of leaving a vehicle and moving towards exits was made and sometimes later, when overcrowded region were formed. Such regions could be found in the vicinity of the exits. There are 3 groups of many overlapped paths visible in all figures. They correspond to the bus passengers, who all were moving together in a similar way. This is the main cause of mentioned above overcrowded regions. In all examined cases people who were trapped inside the tunnel but close to its portals chose their way of escape towards the portals and managed to avoid any contact with the smoke.

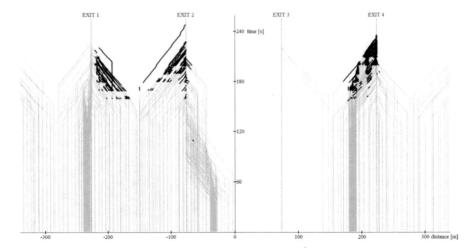

Fig. 7. Paths of the evacuation for case C

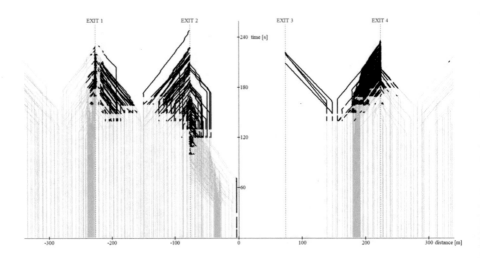

Fig. 8. Paths of the evacuation for case D

As can be seen in Fig. 5, which shows the evacuation process for case A, the majority of the evacuees reached the exits or the portals without any contact with the smoke. According to the simulation results just 70 persons were regarded as endangered due to worsened visibility. It concerns mainly the bus passengers on leeward, who moved towards the 4-th exit.

When analyzing case B (Fig. 6) one can easily notice that people who moved towards exits no 2, 3 and 4 were clearly endangered by the smoke. The total number of the endangered evacuees was 127. The smoke covered them at about 120 s of the fire development, 60 s after the alarm had been triggered. The difference in relation the case A comes from the much higher fire power, which is connected with higher smoke production rate and stronger buoyancy forces. As can be seen many people were caught by the smoke just in the vicinity of exits. It is especially visible for people waiting near the 2-nd exit.

Thanks to quick fire detection the evacuation process in cases A and B was completed in about 180 s. It allowed most of the people to escape before the smoke spread widely in the tunnel.

Cases C and D correspond to the delayed fire detection (120 s). In such circumstances, in case C only the people who were the eyewitnesses of the accident and the fire ignition managed to escape (apart of those who moved toward the tunnel portals). It concerns even the bus passengers. The total number of endangered people was 127, what is a slightly lower value than for case B; however one should has in mind that fire power was here 4 times lower. The smoke covered the evacuees who moved towards exits no 1, 2 and 4 at about 140 s of the fire development. It means that these people had to move at considerable distance being surrounded by smoke, so they might lose their way or even be poisoned by toxic gases.

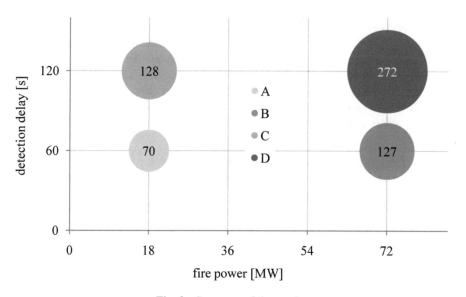

Fig. 9. Summary of the results

Case D was the worst one of the examined. As it can be seen almost all people trapped far from portals were endangered to the smoke influence. The total number of endangered people was 272, what is a clear majority of evacuees. Due to the high fire power the smoke covered the trapped people between 120 and 160 s. It is in accordance with Fig. 4D, where one can see almost all tunnel volume filled by smoke at 150 s of fire development. Even the people who decided to escape by themselves were finally caught by smoke.

Figure 9 present a short summary of the obtained results. The circles area visualizes the number of endangered people. Having in mind, that as was mentioned earlier the total number of trapped inside the tunnel people is 398, it can be clearly seen that even in the least serious case of high power fire the expected number of casualties could be terrifying.

6   Conclusions

The work presented simulation experiments of high power fire development as a consequence of a road accident in an urban road tunnel with congested traffic. Four different fire and evacuation scenarios were examined. The scenarios combined a quick or a delayed fire detection with a medium fire power (a typical bus) or a large fire power (a HGV with a combustible load). The obtained results proved that the number of endangered people would be high even under optimistic assumptions. The situation clearly turned worse for delayed fire detection or large fire power.

The results are alarming despite the tunnel is equipped with 4 emergency exits. The exits are located every 150 m, meanwhile Polish regulations allow the distance to be not greater than 500 m [44]. In other countries the regulations are usually more stringent and commonly the distance of 250–300 m is adopted [8].

A special attention was paid to the evacuation process of bus passengers. As far as a passenger of a car has just to recognize the situation and to make the decision of evacuation and then he is able to escape, bus passengers additionally have to get off their vehicle. Next, they move together in a sizable group, what hinders the evacuation, especially in the close vicinity of emergency exits.

The carried out experiments allowed also for stating the following detailed conclusions:

- the fire detection systems should be able to react quickly,
- the fire detection systems should be multiplied and be sensitive to different fire symptoms; it would assure their reliability and efficiency,
- HGVs with danger load should be prohibited to enter urban tunnels, especially when traffic is congested,
- emergency markings should be visible and clearly guide the evacuees towards safe places even when the visibility is worsened; therefore emergency markings should be located as low as possible - authors suggest the blinking signs mounted directly in the pavement,
- an emergency pattern of the operation of the ventilation system has to keep as large as possible tunnel volume free of smoke; it has to be switched on suitably to fire location and fire power; its design has to take into account the time needed for reaching the full capacity as well [45].

References

1. Kashef, A.Z., Benichou, N.: Investigation of the performance of emergency ventilation strategies in the event of fire in a road tunnel-a case study. J. Fire Prot. Eng. **18**(3), 165–198 (2008)

2. Kumar, S.: Recent achievements in modelling the transport of smoke and toxic gases in tunnel fires. https://about.ita-aites.org/component/k2/tag/Safe%20and%20reliable%20Tunnels?start=10

3. National Fire Protection Association: Standard for Road Tunnels, Bridges, and Other Limited Access Highways 502. An International Codes and Standards Organization, Dallas (2017)

4. Technical Committee 5 Road Tunnels: Fire and smoke control in road tunnels. The World Road Association (1999)

5. Aralt, T.T., Nilsen, A.R.: Automatic fire detection in road traffic tunnels. Tunn. Undergr. Space Technol. **24**, 75–83 (2009)

6. Król, A., Król, M.: Study on hot gases flow in case of fire in a road tunnel. Energies **11**(13), 1–16 (2018)

7. Kashef, A.Z., Viegas, J., Mos, A., Harvey, N.: Proposed idealized design fire curves for road tunnels. https://nrc-publications.canada.ca/eng/view/accepted/?id=e96ebac3-0c6c-4cf8-b4ff-ca9dda9d845c

8. RABT: Forschungsgesellschaft fur Strassen-and Verkehrswesen, Richtlinien fuer Ausstattung und Betrieb von Strassentunneln. https://www.cob.nl/document/richtlinien-fur-die-ausstattung-in-der-betrieb-von-strassentunneln-rabt/

9. Klote, J.H., Milke, J.A., Turnbull, P.G., Kashef, A., Ferreira, M.J.: Handbook of Smoke Control Engineering. ASHRAE, Atlanta (2012)

10. Memorial Tunnel Fire Ventilation Test Program: Test Report, Technical report. Massachusetts Highway Department and Federal Highway Administration, Massachusetts (1995)

11. Lee, K.S., Eom, J.K., Moon, D.: Applications of TRANSIMS in transportation: a Literature review. Procedia Comput. Sci. **32**, 769–773 (2014)

12. Pribyl, O., Pribyl, P., Horak, T.: System for deterministic risk assessment in road tunnels. Procedia Eng. **192**, 336–341 (2017)

13. Nilsson, D., Johansson, M., Frantzich, H.: Evacuation experiment in a road tunnel: a study of human behavior and technical installations. Fire Saf. J. **44**, 458–468 (2009)

14. Canter, D., Breaux, J., Sime, J.: Domestic, multiple occupancy, and hospital fires. In: Canter, D. (ed.) Fires and Human Behavior, pp. 117–136. Wiley, New York (1980)

15. Sime, J.D.: Movement towards the familiar - person and place affiliation in a fire entrapment setting. Environ. Behav. **17**(6), 697–724 (1985)

16. Tong, D., Canter, D.: The decision to evacuate: a study of the motivation which contribute to evacuation in the event of fire. Fire Saf. J. **9**, 257–265 (1985)

17. Proulx, G., Sime, J.D.: To prevent 'panic' in an underground emergency: why not tell people the truth? Fire Saf. Sci. **3**, 843–852 (1991)

18. Frantzich, H.: Occupant behavior and response time - results from evacuation experiments, in Human Behaviour in Fire. https://www.tib.eu/en/search/id/BLCP%3ACN040321500/Occupant-behaviour-and-response-time-Results-from/

19. Nilsson, D., Frantzich, H., Saundres, W.: Colored flashing lights to mark emergency exits - experiences from evacuation experiments. In: Fire Safety Science. Proceedings of the Eighth International Symposium, pp. 569–579 (2005). https://www.iafss.org/publications/fss/8/569/view/fss_8-569.pdf

20. Noren, A., Winer, J.: Modelling crowd evacuation from road and train tunnels - data and design for faster evacuations. Report 5127. Lund University, Department of Fire Safety Engineering, Lund (2003)

21. Nilsson, D., Johansson, A.: Social influence during the initial phase of a fire evacuation - analysis of evacuation experiments in a cinema theatre. Fire Saf. J. **44**(1), 71–79 (2009)

22. Helbing, D., Molnar, P.: Social force model for pedestrian dynamics. Phys. Rev. E Stat. Phys. Plasmas Fluids Relat. Interdiscip. Top. **51**(5), 4282–4286 (1995)

23. Capote, J.A., Alvear, D., Abreu, O., Cuesta, A., Alonso, V.: A real-time stochastic evacuation model for road tunnels. Saf. Sci. **52**, 73–80 (2013)

24. Averill, J.D.: Five grand challenges in pedestrian and evacuation dynamics. In: Peacock, R. D., Kuligowski, E.D., Averill, J.D. (eds.) Pedestrian and Evacuation Dynamics, pp. 1–11. Springer, Boston (2011)

25. Ronchi, E., Nilsson, D., Gwynne, S.M.V.: Modelling the impact of emergency exit signs in tunnels. Fire Technol. **48**, 961–988 (2012)

26. Ronchi, E.: Testing the predictive capabilities of evacuation models for tunnel fire safety analysis. Saf. Sci. **59**, 141–153 (2013)

27. Seike, M., Kawabata, N., Hasegawa, M.: Quantitative assessment method for road tunnel fire safety: development of an evacuation simulation method using CFD-derived smoke behavior. Saf. Sci. **94**, 116–127 (2017)

28. Nawrat, S., Schmidt-Polończyk, N., Napieraj, S.: Safety assessment of road tunnels with longitudinal ventilation, during a fire incident, utilizing numerical modelling tools. Saf. Fire Tech. **43**(3), 253–264 (2016)

29. Shen, Y., Wang, Q., Yan, W., Sun, J., Zhu, K.: Evacuation processes of different genders in different visibility conditions-an experimental study. Procedia Eng. **71**, 65–74 (2014)

30. Smith, J.: Agent-Based Simulation of Human Movements During Emergency Evacuations of Facilities. https://www.wbdg.org/files/pdfs/agent_based_sim_paper.pdf

31. Ronchi, E., Colonna, P., Capote, J., Alvear, D., Berloco, N., Cuesta, A.: The evaluation of different evacuation models for assessing road tunnel safety analysis. Tunn. Undergr. Space Technol. **30**, 74–84 (2012)

32. Ronchi, E., Colonna, P., Berloco, N.: Reviewing Italian Fire Safety Codes for the analysis of road tunnel evacuations: advantages and limitations of using evacuation models. Saf. Sci. **52**, 28–36 (2013)

33. Mu, N., Song, W., Qi, X., Lu, W., Cao, S.: Simulation of evacuation in a twin bore tunnel: analysis of evacuation time and egress selection. Procedia Eng. **71**, 333–342 (2014)

34. General Directorate of National Roads and Motorways. https://www.gddkia.gov.pl/pl/a/21644/Generalny-Pomiar-Ruchu-2015-wyniki

35. British Standards 7346-4: Components for smoke and heat control systems. Functional recommendations and calculation methods for smoke and heat exhaust ventilation systems, employing steady-state design fires. Code of practice (2003)

36. Beard, A., Carvel, R.: The Handbook of Tunnel Fire Safety. Thomas Telford Ltd., London (2005)

37. Lonnermark, A., Ingason, H.: Fire spread and flame length in large-scale tunnel fires. Fire Technol. **42**, 283–302 (2006)

38. Guo, X., Zhang, Q.: Analytical solution, experimental data and CFD simulation for longitudinal tunnel fire ventilation. Tunn. Undergr. Space Technol. **42**, 307–313 (2014)

39. Kashef, A.: Ventilation strategies - an integral part of fire protection systems in modern tunnels. In: 7th International Symposium on Tunnel Safety and Security, pp. 16–18. SP Technical Research Institute of Sweden, Montréal (2016)

40. Węgrzyński, W., Krajewski, G.: Wykorzystanie badań w skali modelowej do weryfikacji obliczeń CFD wentylacji pożarowej w tunelach komunikacyjnych. Budownictwo Górnicze i Tunelowe **4**, 1–7 (2014)

41. PIARC: Design fire characteristics for road tunnels. Norman Rhodes, Paris (2017)

42. Jannsens, M.: Development of a database of full-scale calorimeter tests of motor vehicle burns. Southwest Research Institute, San Antonio Texas (2008)

43. Mikame, Y., Kawabata, N., Seike, M., Hasegawa, M.: Study for safety at a relatively short tunnel when a tunnel fire occurred. https://www.tunnel-graz.at/history/Tunnel_2014_CD/Dateien/19_Mikame_2014_neu.pdf
44. Minister of Transport and Maritime Economy: Regulation of the Minister of Transport and Maritime Economy of 30th May 2000 on technical conditions to be met by road engineering facilities and their location. The Chancellery of the Sejm of the Republic of Poland, Warsaw (2000)
45. Król, A., Król, M.: Transient analyses and energy balance of air flow in road tunnels. Energies **11**(7), 1–15 (2018)

Analysis of Driver Behaviour at Roundabouts in Tokyo and the Tokyo Surroundings

Elżbieta Macioszek$^{(\boxtimes)}$

Faculty of Transport, Silesian University of Technology, Katowice, Poland
elzbieta.macioszek@polsl.pl

Abstract. The article comments upon results of an analysis of driver behaviour at roundabouts in Tokyo and the Tokyo surroundings. This analysis was conducted as a part of a project entitled "Analysis of the applicability of the author's method of roundabouts entry capacity calculation developed for the conditions prevailing in Poland to the conditions prevailing at roundabouts in Tokyo (Japan) and in the Tokyo surroundings" financed by the Polish National Agency for Academic Exchange. The article analyses the patterns of behaviour of drivers at roundabout entry legs with regard to the rule of limited confidence in other drivers at the roundabout and the impact of these behaviour patterns on the entry leg's flow capacity. The results thus obtained show a better match between the models applied for estimating the roundabout entry leg capacity, where one takes into account a certain part of the stream of vehicles leaving the main circulatory roadway using the exit preceding the entry leg analysed, compared to a model where the stream of vehicles leaving the main circulatory roadway has not been included in the composition of the higher rank volume for the given entry leg.

Keywords: Roundabouts · Driver behaviors · Roundabouts entry capacity · Road traffic engineering · Transport

1   Introduction

Despite numerous earthquakes due to the fact that the country is located in a seismically active area, Japan's transport system is technologically advanced and among the most highly developed in the world. It clearly stands out with the high level of quality of passenger and freight transport services. Japan is a left-hand traffic country. The length of its road network is 1,304,700 km, which makes it the fifth largest in the world (after the USA, India, China and Brazil). The road network density index is the highest in the world at 316.8 km/100 km^2, and the network of connections is very well maintained. The rapid growth of the automotive industry has contributed to the development of the toll motorway system linking all major cities, with the total length of 6,114 km. There are 375 passenger cars per 1,000 inhabitants in Japan. This number is relatively low compared to other highly developed countries. However, road transport is not essential to the Japanese passenger and freight transport, as this function is mainly performed by rail transport.

© Springer Nature Switzerland AG 2020
E. Macioszek and G. Sierpiński (Eds.): Modern Traffic Engineering in the System Approach to the Development of Traffic Networks, AISC 1083, pp. 216–227, 2020.
https://doi.org/10.1007/978-3-030-34069-8_17

The year 1872 saw the launching of the first Japanese railway line connecting Tokyo with Yokohama. At the beginning of the 20th century, the first lines within the Tokyo railway junction were electrified, while before the outbreak of World War II, the first metro sections were commissioned in Tokyo and Osaka. Nowadays, Japan has the most highly advanced and comprehensively developed rail transport system in the world, with the total length of its railway lines exceeding 30,000 km. The density of railway lines is 6.21 km/100 km^2. The first Shinkansen railway line (with trains running at a speed of 300 km/h) was launched in 1964 to connect Tokyo with Osaka. At present, the Shinkansen railway line is 3,000 km long, and still being extended. It has been fully electrified and automated. The four largest islands of the archipelago are linked together by railway with the main carrier being the Japan Rail group. The metro operates in such cities as Tokyo, Kyoto, Yokohama, Sapporo, Osaka, Kōbe, Nagoya, Fukuoka and Sendai. Moreover, the network of suburban connections is extended with overhead monorail and tram lines.

Japan offers an excellently developed air transport system which performs an essential role in the international passenger traffic. Japan is only second to the USA in terms of the volume of freight and passenger transport services worldwide. Japan's main international airports are the following:

- Haneda, Tokyo (the largest Japanese airport and the fourth largest airport in the world, serving only domestic flights),
- Narita, Tokyo (International Airport),
- Kansai, Osaka (international airport, and the largest airport built on an artificial island at the same time),
- Chūbu, Nagoya (international airport located on an artificial island near Tokoname).

The main Japanese air carriers are:

- Japan Airlines (JAL),
- Nippon Airways (ANA),
- Skymark Airlines, Skynet Asia Airways (airlines with a major share in the volume of domestic transport services).

In terms of maritime transport, the merchant fleet consists of 702 vessels with a total tonnage of 12,680,544 DWT, being one of the largest in the world, second only to Norway among the most highly developed countries. The fleet comprises tankers, bulk carriers, gas carriers, ferries, RO-RO ships, general cargo carriers, liners, chemical carriers, refrigerated carriers, container carriers and other vessels. As the country is composed of numerous islands, its maritime transport accounts for a major volume of transport services concerning both exported and imported goods. The main shipping routes include those connecting Japan with the USA, Australia, Europe and the Persian Gulf. The largest Japanese seaports include: Nagoya, Chiba, Yokohama, Kitakyushu, Osaka, Tokyo and Kobe. The most important Japanese shipowners are:

- K Line,
- Mitsui O.S.K. Lines,
- Nippon Yusen (NYK Line).

In Poland, as in other European countries, roundabouts are designed and implemented frequently and willingly. A significant number of both domestic and foreign scientific publications (including [1–14]) have already been dedicated to these infrastructural components. In many cases, roundabouts, being elements of the transport network functioning in an area subject to specific studies, were also perceived as research problems of secondary nature, with a different primary objective of the given research (e.g. [15–24]). They also contribute to an environmentally friendly transport system [25]. However, roundabouts are not particularly popular in Japan. A decided majority of road junctions are controlled with traffic lights. The article comments upon results of an analysis of driver behaviour at roundabouts in Tokyo and the Tokyo surroundings. It analyses the patterns of behaviour of drivers at roundabout entry legs with regard to the rule of limited confidence in other drivers at the roundabout and the impact of these behaviour patterns on the entry leg's flow capacity. The results thus obtained show a better match between the models applied for estimating the roundabout entry leg capacity, where one takes into account a certain part of the stream of vehicles leaving the main circulatory roadway using the exit preceding the entry leg analysed, compared to a model where the stream of vehicles leaving the main circulatory roadway has not been included in the composition of the higher rank volume for the given entry leg.

2 Problems of Drivers' Limited Confidence in Other Drivers at Roundabouts

The stream of vehicles on the roundabout's main circulatory roadway between two successive entry legs consists both of vehicles whose drivers intend to leave the main roadway using another exit and of vehicles whose drivers intend to continue driving on the main circulatory roadway.

The volume of the stream of vehicles using the roundabout's main circulatory roadway of a higher rank to the drivers of vehicles using the given entry leg (Q_{nwl}), is typically assumed as a sum of traffic volumes of all routes comprising the stream using the main roadway upstream the given entry leg (less frequently, what one also includes in this volume is the routes which clearly affect the responses of drivers using a minor route, leaving the junction's main circulatory roadway with an exit leg adjacent to the analysed entry leg).

There are papers in the literature of the subject which emphasise that one can observe an interaction between drivers of vehicles on the main circulatory roadway with an intention to leave the main roadway using an exit adjacent to the analysed entry leg and drivers queuing at the entry leg waiting for a possibility to enter the circulatory roadway, and these include such publications as [26–30].

Assuming certain characteristics of the vehicle stream and the roundabout geometry, such reactions between the drivers of vehicles already on the roundabout's main circulatory roadway and those at the start of the exit leg and at the entry leg reduce the entry leg capacity, even though the vehicle drivers leaving the main circulatory roadway are not considered to belong to the stream of a higher rank to the given entry leg. The interactions in question are typically not taken into account in models used to estimate the flow capacity of roundabout entry legs (the exceptions being the French and the Swiss method for calculating the roundabout flow capacity) but only in models enabling the flow capacity of entry legs to be estimated for non-signal-controlled intersections (which is the case of the Polish method, for instance).

On account of the comparable behaviour patterns displayed by drivers turning right from a minor entry leg at simple three-leg intersections and entering from a minor entry leg at roundabouts, one may expect that to a certain degree the effect of the vehicles leaving the roundabout's main circulatory roadway on the entry leg capacity should be similar to that observed at simple intersections. With regard to the said interactions between vehicle drivers, it can be concluded that it is for the specificity of circular vehicle traffic on the main circulatory roadway that the trajectories of vehicles leaving the main roadway can only be clearly distinguished from those of vehicles continuing on the intersection's main roadway at the time when the driver using the main roadway actually begins the manoeuvre of exiting with the selected exit leg.

What probably also exerts a negative impact on the confidence of roundabout users in other drivers is the lack of splitter islands at some entry legs or the fact that the islands have only been developed as pavement markings, and indeed, numerous entry legs in Japan are only in the form of pavement markings (Fig. 1).

Fig. 1. Pavement-marked traffic island dividing the entry and the exit leg at the roundabout in front of JR Hitachi-Taga Station in Hitachi, Ibaraki Prefecture (Japan)

The lack of divisional islands actually increases the influence exerted by the vehicle drivers leaving the roundabout on the acceptance of time intervals on the main roadway by the vehicle drivers at the adjacent entry leg. What may also affect the process of entry by vehicle drivers into the main circulatory roadway is the form in which the central roundabout island has been established. When developed into a low structure (e.g. by concreting or paving), it enables the driver to analyse the traffic situation on the entire main circulatory roadway, i.e. on the other side of the central island as well, and sometimes even at the roundabout's other entry legs, making the driver more inclined to wait for larger and safer intervals in the higher rank stream (example - Fig. 2).

Fig. 2. Example of a low central island of a roundabout in Hanyu, Saitama Prefecture (Japan)

Where the central island has been developed into a higher structure (e.g. using vegetation), which prevents the full overview of the traffic situation on the round-about's roadway, drivers tend to decide more quickly about entering the main circu-latory roadway. Moreover, based on observations of the behaviour of Japanese drivers, one may conclude that compared to Polish drivers the former are more inclined to give right of way, even when they actually have the right of way, in which case they behave very courteously by yielding to other approaching drivers.

Another factor which may also affect the process of entry into the main circulatory roadway is the size of the town/district where the roundabout is located. In large cities, invariably facing traffic problems caused by high levels of congestion in the transport network, drivers tend to move in a more compact and dynamic manner, and make decisions to enter the main roadway more quickly, even if by doing that they should force other drivers moving in the roundabout roadway stream with the right of way to brake. In contrast, drivers from smaller towns display more distrust towards other drivers on roundabouts, and are willing to wait longer at the entry leg until they are fully confident whether an approaching driver on the main circulatory roadway leaves the main roadway or continues on the roundabout.

The lack of confidence in the behaviour of drivers queuing at entry legs is also caused by the limited confidence in vehicle drivers moving on the main roadway, who use direction indicators to signal their intention to exit the main circulatory roadway, turn on the direction indicator too late or do not turn it on at all. It was also often observed in field measurements that direction indicators were used by drivers in a completely incomprehensible manner, or inadequately to the manoeuvre actually being performed.

At certain roundabouts, one can observe that the way in which the area adjacent to the roundabout has been developed is the reason why the angle between two adjacent entry legs carrying the traffic to the intersection is different from 90°. In the event that the angle is less than 90°, the axes of adjacent entry legs are very close to each other, which in certain situations may cause that direction indicators are used at the given intersection in a manner which some traffic participants simply cannot comprehend. The above traffic situation usually pertains to two consecutive roundabout legs located close to each other.

3   Analysis of Driver Behaviour Against the Limited Confidence in Other Drivers at Roundabouts

In light of the above elaborations, based on the research material collected, a survey was conducted at one-lane roundabouts in Tokyo and in Tokyo surroundings to analyse the effect of drivers applying the rule of limited confidence in other road users at roundabout entry legs on the traffic conditions at the entry legs. A preliminary analysis of the survey results thus obtained showed that, on average, six out of ten drivers entering the main circulatory roadway from an entry leg followed the rule of limited confidence in other road users (Fig. 3). In Fig. 3, Z_{wl} designates the number of drivers applying the rule of limited confidence in other traffic participants at entry leg wl, where $wl = 1, 2, 3,..., n.$, over a period of one hour, while Q_{wl} is the traffic volume at entry leg wl for $wl = 1, 2, 3,..., n.$, over a period of one hour.

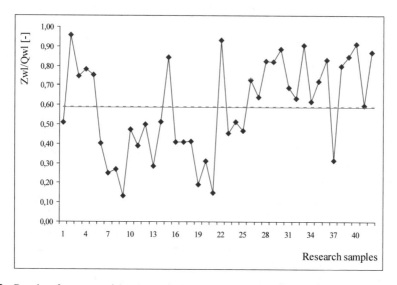

Fig. 3. Results of a survey of the share of drivers applying the rule of limited confidence in other traffic participants at entry legs of one-lane roundabouts in Tokyo and Tokyo surroundings

As a result of an analysis of the empirical data acquired, it was established that in a major part of the examined range of the entry leg capacity, the cumulative frequencies of the empirical entry leg capacity (C_{wIE}) displayed visual convergence with the cumulative distribution function of the entry leg capacity determined by including 75% of the value of the stream leaving the main circulatory roadway in the value of the higher rank traffic volume, i.e. $Q_{nwl} + (0.75 \div Q_{op.obw.wy.})$. The results of this comparison have been presented in Fig. 4.

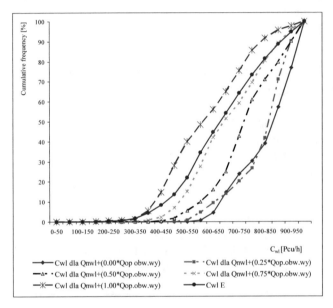

Fig. 4. Cumulative distribution functions of entry leg capacity at one-lane roundabouts for cases of the entry leg capacity estimation by considering the following values in calculations: 0%, 25%, 50%, 75% and 100% of the volume of the vehicle stream leaving the main circulatory roadway with an exit leg adjacent to the entry leg analysed

This convergence was subject to preliminary analysis using the Kolmogorov-Smirnov λ test to verify the zero hypothesis concerning compatibility of the cumulative distribution functions for the distributions analysed: $H_o : F(C_{wl}\ dla\ Q_{nwl} + (0.75 \cdot Q_{op.obw..wy})) = F(C_{wIE})$ at the significance level of α = 0.05, versus the alternative hypothesis assuming that there are significant differences in the forms of the functions analysed: $H_1 : F(C_{wl}\ dla\ Q_{nwl} + (0.75 \cdot Q_{op.obw.wy.})) \neq F(C_{wIE})$. The statistic value thus received ($D = 0.190 < D_{\alpha=0.05} = 0.22$) has confirmed that there are no grounds to reject zero hypothesis H_o. On the other hand, the statistic value of $D = 0.41 > D_{\alpha=0.05} = 0.22$ obtained in the Kolmogorov-Smirnov λ test performed to investigate the zero hypothesis at the level of significance of α = 0.05 concerning conformity between the empirical cumulative distribution function of the entry leg capacity and the cumulative distribution function of the entry leg capacity, without considering the higher rank stream leaving the main circulatory roadway while establishing the flow capacity value

$(Q_{nwl} + (0.00 \cdot Q_{op.obw.wy}))$, has made it possible to reject the zero hypothesis in favour of the alternative hypothesis assuming that there are significant differences in the forms of the cumulative distribution functions analysed. With reference to the above conclusions, one can claim that including a certain part of the stream of vehicles leaving the main circulatory roadway of a one-lane roundabout in the higher rank volume for the given entry leg improves the quality of model matching with empirical data.

This fact is further confirmed by comparing empirical values of entry leg capacity with the flow capacity values calculated for an entry leg using the following models: C_{wl} for $Q_{nwl} + (0.00 \cdot Q_{op.obw.wy})$, C_{wl} for $Q_{nwl} + (0.25 \cdot Q_{op.obw.wy})$, C_{wl} for $Q_{nwl} + (0.50 \cdot Q_{op.obw.wy})$, C_{wl} for $Q_{nwl} + (0.75 \cdot Q_{op.obw.wy})$, C_{wl} for $Q_{nwl} + (1.00 \cdot Q_{op.obw.wy})$. Based on this comparison, it was found that nearly all values of entry leg capacity C_{wl} for $Q_{nwl} + (0.00 \cdot Q_{op.obw.wy})$ were above a line inclined at 45° to the OX axis (representing a perfect match of data), meaning that the entry leg capacity model which does not consider vehicles leaving the main circulatory roadway using an exit leg adjacent to the entry leg subject to analysis returns overestimated values of the entry leg capacity compared to real-life traffic conditions. As a result of the analysis of the quality of model matching with empirical data, determination coefficient $R^2 = 0.43$ [-] was obtained, while the coefficient of determination for the data extracted from the model of entry leg capacity C_{wl} for $Q_{nwl} + (0.25 \cdot Q_{op.obw.wy})$ came to $R^2 = 0.51$ [-]. While the coefficient of determination for the data extracted from the model of entry leg capacity C_{wl} for $Q_{nwl} + (0.25 \cdot Q_{op.obw.wy})$ came to $R^2 = 0.54$ [-], and in the case of the data extracted from the model of entry leg capacity C_{wl} for $Q_{nwl} + (0.50 \cdot Q_{op.obw.wy})$, the coefficient of determination was $R^2 = 0.54$ [-]). The best match was obtained upon comparing the values of empirical entry leg capacity with the entry leg capacity values calculated using model C_{wl} for $Q_{nwl} + (0.75 \cdot Q_{op.obw.wy})$ - Fig. 5. In this case, the

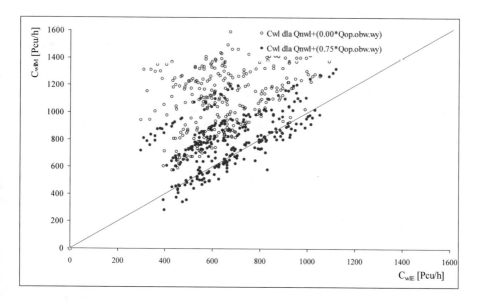

Fig. 5. Comparison of flow capacity values of a roundabout entry leg for $Q_{nwl} + (0.00 \cdot Q_{op.obw.})$ with those of an entry leg for $Q_{nwl} + (0.75 \cdot Q_{op.obw.})$

calculated values are distributed fairly regularly, i.e. above and below the straight line representing the perfect match (the coefficient of determination being $R^2 = 0.71$ [-]).

The results thus obtained show a better match between the models applied for estimating the roundabout entry leg capacity, where one takes into account a certain part of the stream of vehicles leaving the main circulatory roadway using the exit preceding the entry leg analysed, compared to a model where the stream of vehicles leaving the main circulatory roadway has not been included in the composition of the higher rank volume for the given entry leg.

Based on observations of vehicular traffic streams at roundabouts, one may conclude that the value of flow capacity of a roundabout entry leg considered from the perspective of the rule of limited confidence in other drivers at roundabouts depends on factors such as:

- distance between a conventional point in the main roadway, where the trajectories of vehicles leaving the main circulatory roadway are explicitly distinguished by drivers at an entry leg from the trajectories of vehicles continuing on the main circulatory roadway, and a conventional point where the trajectories of the vehicles continuing on the main circulatory roadway intersect with the trajectory of the vehicles entering the main circulatory roadway from an entry leg,
- volume of the stream of vehicles leaving the main circulatory roadway with an exit leg preceding the entry leg subject to analysis,
- speed of the vehicles running on the main circulatory roadway,
- other factors.

The aspects analysed further on during the studies included the values of the relative errors in estimating the flow capacity values for roundabout entry legs (the relative error value was derived from the following formula: $\delta_C \approx \frac{(C_{wl}-C_{wlE})}{C_{wlE}} \cdot 100\%$, where C_{wl} is the value of the calculated entry leg capacity, and C_{wlE} is the empirical entry leg capacity) depending on the distance between a conventional point in the main roadway, where the trajectories of vehicles leaving the main circulatory roadway are explicitly distinguished by drivers at an entry leg from the trajectories of vehicles continuing on the main circulatory roadway, and a conventional point where the trajectories of the vehicles continuing on the main circulatory roadway intersect with the trajectory of the vehicles entering the main circulatory roadway from an entry leg. Having performed these analyses, one could establish the ranges for which one would obtain the smallest values of errors in estimating the entry leg capacity (assuming that the perfect match is the case of $\delta_C \approx 0.00\%$):

$$\delta_C \rightarrow \min \Rightarrow \begin{cases} O_{wl} \leq 10 \ m & gdy \quad Q_{nwl} + \left(1.00 \cdot Q_{op.obw.wy}\right) \\ O_{wl} \in (10; 11\rangle & gdy \quad Q_{nwl} + \left(0.75 \cdot Q_{op.obw.wy}\right) \\ O_{wl} \in (11; 12\rangle & gdy \quad Q_{nwl} + \left(0.50 \cdot Q_{op.obw.wy}\right) \\ O_{wl} \in (12; 13\rangle & gdy \quad Q_{nwl} + \left(0.25 \cdot Q_{op.obw.wy}\right) \\ O_{wl} > 14 & gdy \quad Q_{nwl} + \left(0.00 \cdot Q_{op.obw.wy}\right) \end{cases} \quad (1)$$

4   Conclusions

This article addresses the problem of limited confidence of drivers at roundabout entry legs in drivers of vehicles running on the main circulatory roadway. Based on the research in question, it has been established that there is a certain interaction between drivers at an entry leg and drivers of vehicles leaving the roundabout's main circulatory roadway with an exit leg preceding the entry leg examined. Another analysed aspect was the effect of such behaviour patterns on the flow capacity value of the given entry leg. The results thus obtained show a better match between the models applied for estimating the roundabout entry leg capacity, where one takes into account a certain part of the stream of vehicles leaving the main circulatory roadway using the exit preceding the entry leg analysed, compared to a model where the stream of vehicles leaving the main circulatory roadway has not been included in the composition of the higher rank volume for the given entry leg.

Acknowledgements. The present research has been financed from the means of the Polish National Agency for Academic Exchange as a part of the project within the scope of Bekker Programme "Analysis of the applicability of the author's method of roundabouts entry capacity calculation developed for the conditions prevailing in Poland to the conditions prevailing at roundabouts in Tokyo (Japan) and in the Tokyo surroundings".

References

1. Pilko, H., Saric, Z., Zovak, G.: Turbo roundabouts: a brief safety, efficiency and geometry design review. In: Macioszek, E., Akçelik, R., Sierpiński, G. (eds.) Roundabouts as Safe and Modern Solutions in Transport Networks and Systems. LNNS, vol. 52, pp. 3–12. Springer, Cham (2019)
2. Giuffre, O., Grana, A., Tumminello, M.L., Giuffre, T., Trubia, S.: Surrogate measures of safety at roundabouts in AIMSUN and VISSIM environment. In: Macioszek, E., Akçelik, R., Sierpiński, G. (eds.) Roundabouts as Safe and Modern Solutions in Transport Networks and Systems. LNNS, vol. 52, pp. 53–64. Springer, Cham (2019)
3. Fernandes, P., Coelho, M.: Making compact two-lane roundabouts effective for vulnerable road users: an assessment of transport-related externalities. In: Macioszek, E., Akçelik, R., Sierpiński, G. (eds.) Roundabouts as Safe and Modern Solutions in Transport Networks and Systems. LNNS, vol. 52, pp. 99–111. Springer, Cham (2019)
4. Kang, N., Terebe, S.: Estimating roundabout delay considering pedestrian impact. In: Macioszek, E., Akçelik, R., Sierpiński, G. (eds.) Roundabouts as Safe and Modern Solutions in Transport Networks and Systems. LNNS, vol. 52, pp. 112–123. Springer, Cham (2019)
5. Małecki, K.: The roundabout micro-simulator based on the cellular automata model. In: Sierpiński, G. (ed.) Advanced Solutions of Transport Systems for Growing Mobility. AISC, vol. 631, pp. 40–49. Springer, Cham (2018)
6. Macioszek, E.: The application of HCM 2010 in the determination of capacity of traffic lanes at turbo roundabout entries. Transp. Probl. **11**(3), 77–89 (2016)
7. Małecki, K., Wątróbski, J.: Cellular automaton to study the impact of changes in Traffic rules in a roundabout: a preliminary approach. Appl. Sci. **7**(7), Article Number: UNSP 742 (2017)

8. Małecki, K.: The use of heterogeneous cellular automata to study the capacity of the roundabout. In: Rutkowski, L., Korytkowski, M., Scherer, R., Tadeusiewicz, R., Zadech, L. A., Zurada, J.M. (eds.) Artificial Intelligence and Soft Computing. LNAI, vol. 10246, pp. 308–317. Springer, Cham (2017)

9. Turoń, K., Czech, P.: Study on roundabouts in Polish conditions - law, safety problems, sanctions. In: Macioszek, E., Akçelik, R., Sierpiński, G. (eds.) Roundabouts as Safe and Modern Solutions in Transport Networks and Systems. LNNS, vol. 52, pp. 137–146. Springer, Cham (2019)

10. Macioszek, E.: The comparison of models for follow-up headway at roundabouts. In: Macioszek, E., Sierpiński, G. (eds.) Recent Advances in Traffic Engineering for Transport Networks and Systems. LNNS, vol. 21, pp. 16–26. Springer, Cham (2018)

11. Turoń, K., Czech, P., Juzek, M.: The concept of walkable city as an alternative form of urban mobility. Scient. J. of SUT **95**, 223–230 (2017)

12. Turoń, K., Golba, D., Czech, P.: The analysis of progress CSR good practices areas in logistic companies based on reports "Responsible Business in Poland. Good Practices" in 2010–2014. Sci. J. SUT **89**, 163–171 (2015)

13. Macioszek, E.: The comparison of models for critical headways estimation at roundabouts. In: Macioszek, E., Sierpiński, G. (eds.) Contemporary Challenges of Transport Systems and Traffic Engineering. LNNS, vol. 2, pp. 205–219. Springer, Cham (2017)

14. Celiński, I., Sierpiński, G.: Method of assessing vehicle motion trajectory at one-lane roundabouts using VIsual techniques. In: Macioszek, E., Akçelik, R., Sierpiński, G. (eds.) Roundabouts as Safe and Modern Solutions in Transport Networks and Systems. LNNS, vol. 52, pp. 24–39. Springer, Cham (2019)

15. Celiński, I.: Evaluation method of impact between road traffic in a traffic control area and in its surroundings. Ph.D. dissertation, Warsaw University of Technology Publishing House, Warsaw (2018)

16. Celiński, I.: GT planner used as a tool for sustainable development of transport infrastructure. In: Suchanek, M. (ed.) New Research Trends in Transport Sustainability and Innovation. SPBE, pp. 15–27. Springer, Cham (2018)

17. Sierpiński, G., Staniek, M., Celiński, I.: Travel behavior profiling using a trip planner. Transp. Res. Procedia **14C**, 1743–1752 (2016)

18. Sierpiński, G.: Distance and frequency of travels made with selected means of transport - a case study for the Upper Silesian conurbation (Poland). In: Sierpiński, G. (ed.) Intelligent Transport Systems and Travel Behaviour. AISC, vol. 505, pp. 75–85. Springer, Cham (2017)

19. Sierpiński, G., Staniek, M.: Education by access to visual information - methodology of moulding behaviour based on international research project experiences. In: Gómez Chova, L., López Martínez, A., Candel Torres I. (eds.) ICERI 2016 Proceedings: 9th International Conference of Education, Research and Innovation, pp. 6724–6729. IATED Academy, Valencia (2016)

20. Sierpiński, G., Staniek, M., Celiński, I.: Shaping environmental friendly behaviour in transport of goods - new tool and education. In: Gómez Chova, L., López Martínez, A., Candel Torres I. (eds.) ICERI 2015 Proceedings: 8th International Conference of Education, Research and Innovation, pp. 118–123. IATED Academy, Valencia (2015)

21. Staniek, M., Czech, P.: Self-correcting neural network in road pavement diagnostics. Autom. Constr. **96**, 75–87 (2018)

22. Chmielewski, J.: Impact of transport zone number in simulation models on cost-benefit analysis results in transport investments. IOP Conf. Ser. Mater. Sci. Eng. **245**, 1–10 (2017)

23. Chmielewski, J.: Transport demand model management system. IOP Conf. Ser. Mater. Sci. Eng. **471**, 1–10 (2019)

24. Chmielewski, J., Olenkowicz-Trempała, P.: Analysis of selected types of transport behaviour of urban and rural population in the light of surveys. In: Macioszek, E., Sierpiński, G. (eds.) Recent Advances in Traffic Engineering for Transport Networks and Systems. LNNS, vol. 21, pp. 27–36. Springer, Cham (2018)
25. Jacyna, M., Żak, J., Jacyna-Gołda, I., Merkisz, J., Merkisz-Guranowska, A., Pielucha, J.: Selected aspects of the model of proecological transport system. J. KONES Powertrain Transp. **20**, 193–202 (2013)
26. Greibe, P., Lund, B.: Capacity of 2-Lane Roundabouts. Trafitec, Denmark (2009)
27. Berhanu, G., Blakstad, F.: Capacity Evaluation of Roundabout Junctions in Addis Ababa. Addis Abeba University, Addis Abeba (2007)
28. Linse, L.: Capcal for Small Roundabouts. Current Status and Improvements. Linkoping University, Linkoping (2010)
29. Mereszczak, Y., Dixon, M., Kyte, M., Rodegerdts, L.A., Blogg, M.: Incorporating Exiting Vehicles in Capacity Estimation at Single-lane U.S. Roundabouts. http://citeseerx.ist.psu.edu/viewdoc/download?doi=10.1.1.508.7303&rep=rep1&type=pdf
30. Oh, H., Sisiopku, V.: Probabilistic models for pedestrian capacity and delay at roundabouts. In: Transportation Research Circular E-C018. Proceedings of the 4th International Symposium on Highway Capacity, pp. 459–470. Transportation Research Board, Hawai (2000)

Structure and Traffic Organization
in Transport Systems

Impact of Sunday Trade Ban
on Traffic Volumes

Jacek Chmielewski[(⊠)]

Faculty of Construction, Architecture and Environmental Engineering,
University of Technology and Life Sciences, Bydgoszcz, Poland
{jacek-ch, zikwb}@utp.edu.pl

Abstract. The results of the analysis of the impact of the trade ban on Sundays into the daily traffic volumes on weekend days on the road network of the medium Polish city are presented in this paper. The analyzes were carried out on the basis of data collected by the local ITS system, on traffic volumes counted in a continuous mode. Comparison of the results of traffic volume counts for the whole 2018 year, in which every two Sundays per month were indicated as allowed for trade, enabled a large research sample to determine changes in these volumes. The presented data may provide guidance for future activities in the field of planning and organizing road transport on weekend days, as well as enable determination of the impact of trade restrictions on transport pollution and environment.

Keywords: Transport · Traffic volume · Environment · Trade ban

1 Introduction

Traffic analyzes for weekend periods are not a frequent object of interest for researchers. They deal with it mainly in case of road works that may disturb a typical weekend traffic or in case of huge events like music festivals or sports holidays. Usually traffic volumes on these days are lower than those observed in typical work days, and the residents' trip destinations are mostly associated with resting and shopping activity. Also, these days are not an area of analysis required for feasibility studies [1], and the typical road network in most of urban areas usually offers an adequate level of traffic conditions for its users. Meanwhile, traffic that is a source of noise and pollution emission [2, 3] significantly reduces the comfort of weekend resting activity and can impact of a residents stress.

So called trade Sundays [4], where trade activity is not banned, are over 20 years of tradition in Poland. By 2018, shops and malls were allowed to be opened for customers on all Sundays except public holidays, Easter and Christmas time. This legislation was mainly used by large shopping centers being visited by crowds of customers. Shopping centers, known also as malls, understood as objects with a retail area exceeding 10.500 sqm, have already been built in most Polish cities. Over 25 such objects were built in Poland in 2018, giving 365 thousand sqm. Of new retail space, and nowadays there are 450 large shopping centers in the whole country [5]. The very good prospects for Poland's economic development make the growth potential of the trade market big, and

© Springer Nature Switzerland AG 2020
E. Macioszek and G. Sierpiński (Eds.): Modern Traffic Engineering in the System Approach to the Development of Traffic Networks, AISC 1083, pp. 231–241, 2020.
https://doi.org/10.1007/978-3-030-34069-8_18

it depends on the owners and managers of shopping centers to what extent it will translate into an increase in the value of their portfolios. Despite the trade ban in the selected Sundays have being implemented since March 2018, the consolidation of the commercial property market is progressing. In the largest cities of Poland, there are over 1.100 square meters on average of a trade space per 1.000 inhabitants. This clearly demonstrates the role shopping malls play in Polish cities. They are both a place for shopping, meetings, entertainment, but also work places for residents. A huge number of employees are employed in shopping malls, without which they cannot function in proper way. Undoubtedly Sunday, like Saturday, it is the only day in the week when residents could calmly fulfil their shopping demands and spend their free time. Hence the great interest if residents in this type of objects. The natural consequence of the functioning of shopping centers on weekends is the increased car traffic. Shopping centers, usually with a large number of free of charge parking places, are often visited by the motorized part of citizens. At the same time, the potential purchases made at these centers are not conducive to travel by alternative forms of transport, such as a public transport or a bicycle. Hence, high traffic volumes in the areas of shopping centers in Poland have not been a new phenomenon on cities' road network. This applies especially to pre-holiday periods, post-holiday periods, promotion and sales periods or unfavourable weather conditions.

The gradual trade ban on Sundays, that has being implemented since March 2018 in Poland translates into many aspects of everyday life, primarily for residents of medium and large cities and their suburbs. This ban did not include the first and last Sunday of the month in 2018. In 2019, this ban has been extended and currently it does not apply only to the last Sunday of the month. According to the government's plans, in 2020 this ban will not cover only selected Sundays. This means that the transitional period, which makes it possible to research the impact of trade Sundays on road traffic, is very short.

The issues of research and analysis of road traffic volumes have already been the subject of many scientific studies, including [6–8]. According to research result presented in [9], this article presents the results of road traffic analyses for trade Sundays and Sundays with the trade ban, as well as the preceding Saturdays. The presented results concern the year 2018 since March to December, that is the period when the legislation has been implemented. The purpose of this article is to indicate the impact of trade Sundays on road traffic at weekends, as well as the scale of changes in traffic volumes, depending on the location of the research area in relation to the city center.

2   Study Area

Bydgoszcz, a medium-sized Polish city with an area of 176 km^2, inhabited by 352 thousand inhabitants, and a population density of over 2,000 people per km^2 was accepted as a study area. There are several large shopping centers located in the city, and the youngest of them with a total retail and service area of 103 000 sqm and offering free of charge parking places for over 1.200 vehicles was completed in December 2015. Moreover there are many smaller grocery stores in the city, among others well known in Europe LIDL and ALDI. Those retails are located in various parts of the city and they are easily accessible from the city's main road network.

The ITS system, which includes among the others the traffic control devices based on the Australian SCATS system [10] and directions for alternative routes for road users, has been implement since 2015 in the city. Main arteries are of the city are included in it, and the next streets will be involved in it in the nearest future. The whole solution uses primarily the infrastructure of vehicle detection sensors at the intersection areas, and especially the induction loops located both at inlets and outlets of intersections. Data from these loops are the basis for traffic control and updating traffic signals schemas, but they are additionally collected in a dedicated ITS database for their further use for a traffic management and prediction. The basic road network of the city with the location of selected vehicle detection stations (count locations) is presented in Fig. 1. A total of 158 count locations serviced by 361 detectors were selected for the analysis, for which the results of traffic counts in the full observation cycle were collected in the database, that is since January 2017 to February 2019. The count locations for which measurement errors, failures or incomplete measurement data were detected, have been eliminated from the future work. The road infrastructure administrator is responsible for the correct functioning of the above mentioned detectors, and each detector failure is automatically marked in the system. Thanks to such monitoring system, it is possible to guarantee quick intervention in case of devices failure, and at the same time it is quit easy to detect any disruption in their functioning.

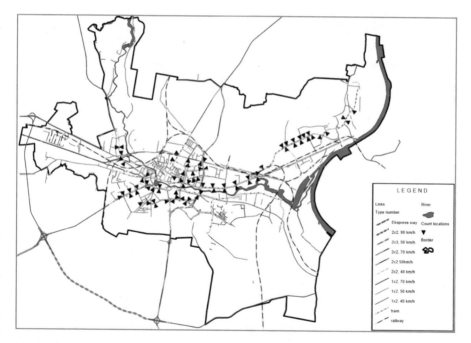

Fig. 1. The basic road network of the city with the location of selected vehicle detection stations (count locations)

For each of count location the distance from the count location to the city center was defined. This is understood as the distance of trip by private car in the city's road network from each of them to one selected intersection located in the central part of the study area. These distances have been divided into statistical classes, and their numbers are presented in Fig. 2. The largest number of observation points was located in the distance between 1.5–2.5 km from the city center, as well as between 7.5–10 km, which results from the fact the eastern district of the city inhabited by 20% of the entire population of the study area is located relatively far away from the city centre.

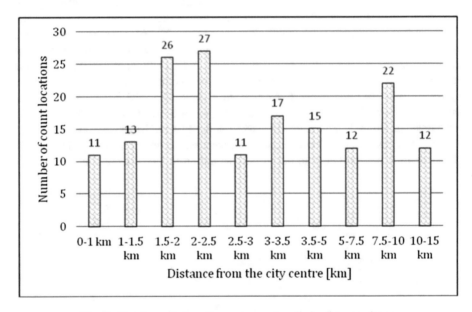

Fig. 2. Numbers of shopping centres and malls in distance classes

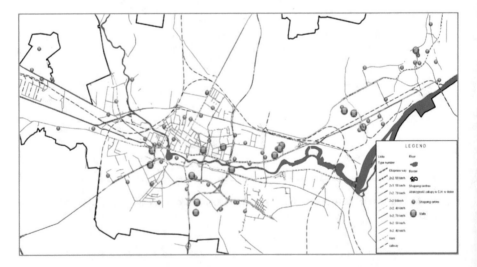

Fig. 3. Location of shopping centres and malls in the study area

The division of count locations into the distance classes allows to determine changes in traffic volume in correlation to the location of a given city area. Theoretically, the closer to the center of the study area, the changes in traffic volumes between trade and no trade days should be higher. As a rule, there is a natural gravity of traffic flow towards the city center, where the most work places, trade and service facilities are located, including shopping centers, restaurants and cafes, pubs, cinemas, parks etc.

In the aspect of trade activity observed in the study area, a total of 76 large-scale retail facilities (with a sales area exceeding 400 m^2) were identified, including 15 malls with an area of over 10,500 m^2. Their location on the background of the city transport network is shown in Fig. 3. All these objects were closed on Sundays marked as with the trade ban.

As it is shown on the presented figure, these objects are located in the area of the entire study area, and their dislocation depends to a large extent on the so-called catchment area - understood as a potential impact force of customers acquisition. There are therefore no single places of special concentration of the above objects. This promotes a large dispersed trips. According to the transportation behaviour research of residents conducted in the study area for weekdays [11], 2.1% of trips from one shopping center is done to the next shopping center.

3 Test Sample and Results of Analyzes

For the purpose of a detailed comparative analysis of traffic volumes at the count locations for Sundays with the trade ban and without the ban, first of all, the data from 2018 were analyzed, in which the number of observation days was similar for each of them. In total, there were 22 Sundays with the trade ban since March to December 2018 (the Easter Sunday was omitted) and for these days a total of 3652 daily traffic counts were collected, and there were 21 Sundays without trade ban with 3486 observations. For these two data sets, average daily traffic volumes were determined and compared. Table 1 shows an exemplary data collected for a single count location.

Table 1. Results of volume comparison for the selected count location.

Month	Trade Sundays	Sundays with the trade ban	Difference [%]
March	6236	4940	26.2
April	6457	5559	16.2
May	6337	5775	9.7
June	6116	5169	18.3
July	5731	4913	16.6
August	5895	4662	26.5
September	6611	5146	28.5
October	6602	5684	16.2
November	6392	4773	33.9
December	6610	5005	32.1

Then all the results were summarized in the table to determine the average changes in daily traffic volumes (DTV) for the entire transport network of the study area covered by the analysis. The results are presented in Table 2 and Fig. 4. As it can be seen from the presented comparison and figure, the increase in traffic volumes on trade Sundays, compared to Sundays with the trade ban, varies from month to month and varies from 12.2% in May up to 30.3% in September, with the weighted average increase in traffic of 22.0%. Such monthly volatility can be explained by weather conditions. The spring and summer months are conducive to recreational activities, and thus the residents abandon shopping's for rest. This way the impact of weather into residents mobility can be noticed.

Table 2. Results of traffic volume comparison for all count locations - Sundays 2018.

Month	Trade Sundays [DTV/day]	Sundays with the trade ban [DTV/day]	Difference [%]
March	7309	5977	22.3
April	7601	6208	22.4
May	7670	6836	12.2
June	7424	6395	16.1
July	7095	5962	19.0
August	7194	5935	21.2
September	7974	6122	30.3
October	8099	6688	21.1
November	7591	5862	29.5
December	7447	5809	28.2

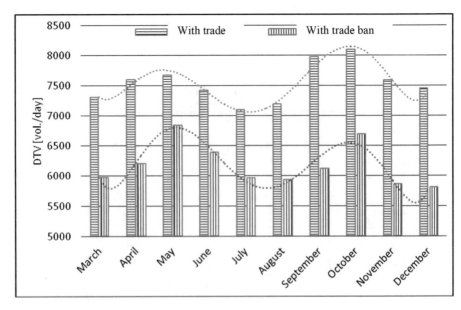

Fig. 4. Results of traffic volume comparison for all count locations and given months Sundays 2018

Thus, in the months since May to July, the increase in traffic volumes is the lowest. August, and especially September are usually months of both: mixed weather conditions and post-season sales as well as the beginning of the school year. These are therefore months of favorable mass shopping activity. And in December, the largest, as much as 30% increase in traffic volumes on the observed transport network was observed. Similarly, in November and December, as months of holidays, especially Christmas time, they were characterized by a high, nearly 30% increase in traffic volumes in trade Sunday compared to traffic volumes on Sundays with the trade ban.

Undoubtedly, the question seems to be whether the Saturdays preceding Sundays with the trade ban are not characterized by increased car traffic and more trips done by the inhabitants of the study area. It seems natural that some residents, considering the limitations in shopping possibilities, decide to meet their typical demands earlier. Therefore, it was necessary to compare the results of road traffic counts at the count locations for Saturdays preceding the analyzed Sundays. The results of such an analysis are presented in Table 3 and Fig. 5. As it can be noticed at the presented statement, the results can be considered quite surprising. In March 2018, i.e. the first month of introducing Sundays with the trade ban, the average daily traffic volumes before such Sunday were lower comparing to the average daily traffic volumes on Saturdays preceding trade Sundays. It is only in the following months that a gradual increase in traffic on Saturdays preceding Sundays with the trade ban without can be observed from nearly 4% up to 8.1% with the average from the entire analysis period at the level of 4.1%. September and October are the exceptions months, in which the changes in average daily traffic volumes between trade Saturdays and Sundays with the trade ban are negligible.

Table 3. Results of traffic volume comparison for all count locations - Saturdays 2018.

Month	Before trade Sundays [DTV/day]	Before Sundays with the trade ban [DTV/day]	Difference [DTV/day]
March	9110	8591	6.0
April	9728	10104	−3.7
May	9228	9992	−7.7
June	8912	9479	−6.0
July	7973	8679	−8.1
August	8359	8583	−2.6
September	9614	9598	0.2
October	9885	9898	−0.1
November	9215	10078	−8.6
December	9314	10142	−8.2

At the same time, it should be underline that the obtained results are supported by a large research data sample, and therefore they should not be treated as a random nature. Once again, the weather conditions may be an explanation for the obtained results. As it comes from historical metrological data all Saturdays in September and October

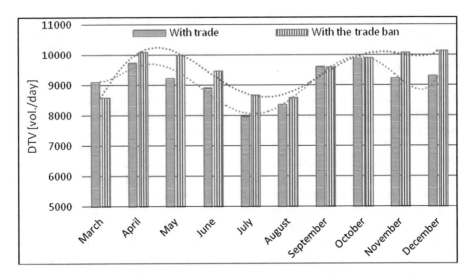

Fig. 5. Results of traffic volume comparison for all count locations and given months - Saturdays 2018

preceding the Sundays with the trade ban were characterized by very good weather conditions conducive to outdoor recreation. The September's sunny Saturdays with temperatures of 26 and 17 degrees of Celsius, as well as October 19 and 10 degrees of Celsius were not conducive to shopping activity.

In order to determine the traffic effects of Sundays without trade ban, Saturday's and Sunday's differences in average daily traffic volumes on trade Sundays and Sundays with the trade ban were balanced. The results are summarized in Tables 4 and 5. As it is observed from the statement, Sundays without the trade ban influence the growth of the overall weekend traffic. The average increase in traffic was estimated at 6.2%, which means 550 vehicles a day in one direction per cross-section. The increase in traffic clearly translates into an increase in noise and pollution emission.

In addition, the spatial distribution of changes in average daily traffic volumes was analysed. For individual groups of traffic count locations (see Fig. 2), the percentage changes in average daily traffic volumes on trade Sundays and Sundays with the trade ban depending on the month were evaluated. The results of the analysis are presented graphically in Fig. 6. As it is presented in the graph presented, the largest changes in averaged daily traffic volumes between trade Sundays and Sundays with the trade ban are observed in the areas farthest away from the city center, that is in compartments above 7 km.

This may be due to the fact that the more central areas are characterized by greater attractiveness to travel on days with the trade ban. There are in fact much more restaurants, cafes, clubs, cinemas as well as parks and other recreational facilities open both on trade Sundays and those with the trade ban. Thus, the differences between daily traffic volumes measured in areas located closer to the central part of the analysis area are lower than those observed at other count locations.

Table 4. Difference in average daily traffic in Saturdays before Sundays with trade and without trade and in Sundays with and without trade.

Month	Difference in DTV Saturdays before Sundays with trade and without trade [DTV/day]	Difference in DTV Sundays with trade and without trade [DTV/day]	Difference total [DTV]
March	519	1333	1852
April	−377	1393	1016
May	−765	833	69
June	−567	1029	462
July	−706	1133	427
August	−224	1259	1035
September	16	1852	1868
October	−13	1411	1398
November	−863	1728	865
December	−829	1638	809

Table 5. Average daily traffic in weekends with Sundays with and without trade.

Month	DTV in weekend trade Sundays [DTV]	DTV in weekend Sundays with the trade ban [DTV]	Difference [%]
March	8210	7284	12.7
April	8664	8156	6.2
May	8449	8414	0.4
June	8168	7937	2.9
July	7534	7320	2.9
August	7776	7259	7.1
September	8794	7860	11.9
October	8992	8293	8.4
November	8403	7970	5.4
December	8380	7976	5.1
		Average	**6.2**

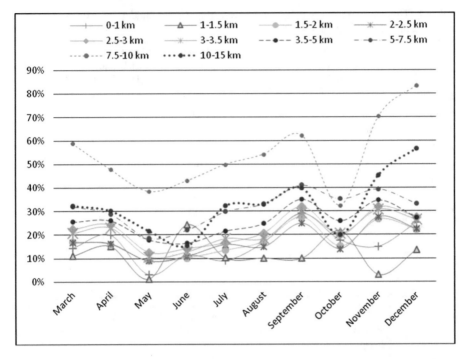

Fig. 6. Results of daily traffic volume changes in months for classes of count locations

4   Conclusion

The results of analyses regarding daily traffic volumes measured on the medium-sized city road network on weekend days in 2018 presented in this paper made it possible to determine the impact of days with trade activity on changes in daily traffic volumes. It has been shown that trade Sundays affect over 6% increase in daily traffic volumes on weekend days in the study area, while traffic volumes on Sunday increase by nearly 20% on average, and at the same time fall by more than 4% on Saturdays that precede them. The most-significant increases in traffic volumes were observed in areas located relatively far away from the city center, with low transport attractiveness on Sundays with the trade ban. In addition, the influence of weather conditions on differences in daily traffic volume on weekend days was observed. Good weather conditions conducive to active rest result in less interest in shopping activity, and thus a smaller increase in traffic volumes compared to days with the trade ban.

The results of the presented research may provide guidelines for cities' authorities, legislation and town planners in the field of mobility policy. Knowledge of the effects of restrictions on trade activity, for example, can be one of the important point of education in the field of conscious mobility and its impact on the comfort of human activity and life.

References

1. JASPERS - Joint Assistance to Support Projects in European Regions, Blue Book (2015). http://www.rpo.malopolska.pl/download/program-regionalny/o-programie/zapoznaj-sie-z-prawem-i-dokumentami/niebieska-ksiega-dla-projektow-infrastruktury-drogowej-realizowa-nych-w-ramach-perspektywy-finansowej-2014–2020/2015/08/Blue_Book_Roads_EN_PL_28052015.pdf
2. Colvile, R.N., Hutchinson, E.J., Mindell, J.S., Warren, R.F.: The transport sector as a source of air pollution. Atmos. Environ. **35**(9), 1537–1565 (2001)
3. Jones, A.M., Yin, J., Harrison, R.M.: The weekday-weekend difference and the estimation of the non-vehicle contributions to the urban increment of airborne particulate matte. Atmos. Environ. **42**(19), 4419–4810 (2008)
4. Balmain Asset Management, BSC Real Estate Advisors: Sunday Trade Ban in Poland. Maximising the trading week, bsc-bam-sunday-ban-market-report 271
5. Cushman & Wakefield: Record Investment Activity on the Polish Commercial Real Estate Market. Cushman & Wakefield, Warsaw (2019)
6. Macioszek, E., Lach, D.: Analysis of the results of general traffic measurements in the west Pomeranian Voivodeship from 2005 to 2015. Sci. J. Silesian Univ. Technol. Ser. Transport **97**, 93–104 (2017)
7. Macioszek, E., Lach, D.: Analysis of traffic conditions at the Brzezinska and Now-ochrzanowska intersection in Myslowice (Silesian Province, Poland). Sci. J. Silesian Univ. Technol. Ser. Transp. **98**, 81–88 (2018)
8. Macioszek, E., Lach, D.: Comparative analysis of the results of general traffic measurements for the Silesian Voivodeship and Poland. Sci. J. Silesian Univ. Technol. Ser. Transp. **100**, 105–113 (2018)
9. Loopik, K.: The Impact of Sunday Shopping Policy on the Dutch Retail Structure. Wageningen University, Wagenigen (2014)
10. Chong-White, C.: The History of SCATS Performance: Information Paper with Peer Reviewed References Information paper. Roads and Maritime Services, New South Wales (2012)
11. Chmielewski, J., Olenkowicz-Trempała, P.: Analysis of selected types of transport behavior of urban and rural population in the light of surveys. In: Macioszek, E., Sierpiński, G. (eds.) Recent Advances in Traffic Engineering for Transport Networks and System. LNNS, vol. 21, pp. 27–36. Springer, Heidelberg (2018)

Analysis of Critical Gap Times and Follow-Up Times at Selected, Median, Uncontrolled T-Intersections Differentiated by the Nature of the Surrounding

Adrian Barchański[✉]

Faculty of Transport, Silesian University of Technology in Gliwice,
Katowice, Poland
adrian.barchanski@polsl.pl

Abstract. The paper presents the results of critical gap times and follow-up times studies at selected median, uncontrolled T-intersections located in the Upper Silesian agglomeration. In the world literature available to the general public, little space was devoted to objects of this type, despite their significant differences in comparison with intersections such as: intersection of 1×2-type roads or median intersection with four approaches. Two selected, real research objects have been characterized. The processes of measurements, data processing and their influence on the sample size was described. During the analysis of the results, attention was drawn to the influence of the selected factors on the critical gap times and follow-up times, in particular on the differential in corresponding relations between intersections within and outside built-up areas. The possibility of using these times for the calibration of capacity calculation methods was pointed out.

Keywords: Critical gap time · Follow-up time · Uncontrolled intersection · Gap determining factors · Siegloch method

1 Introduction

In the dense road network of the city, the time of driving through individual sections and point objects determines the efficiency of the whole system and determines which routes will be most heavily congested, selected by the largest group of users who minimize their travel time in an individual way. Therefore, the planning of traffic flows in the city strictly depends on a thorough analysis of the capacity of individual objects [1–3]. The most frequently used methods of calculating intersection capacity, such as the HCM reference method, the German HBS method with its slightly different approach and selected other methods, make the measures to assess the traffic situation at intersections and capacity estimation dependent on the critical gap times and follow-up times [3–5]. The indicated variables together with the conflicting flow rate are the basic parameters and calibration tool used to determine the capacity of selected intersections involving the mass handling theory model. It is very important that the values of these variables are up to date, in line with reality and reliably reflect the actual

© Springer Nature Switzerland AG 2020
E. Macioszek and G. Sierpiński (Eds.): Modern Traffic Engineering in the System Approach to the Development of Traffic Networks, AISC 1083, pp. 242–256, 2020.
https://doi.org/10.1007/978-3-030-34069-8_19

behaviour of drivers, which are looking forward on minor-street approach. The assessment of traffic conditions at intersections forms the basis for spatial planning, management with regard to infrastructure objects and strategic or operational decision making, and is determined, among others, on the basis of critical gap times and follow-up times values [3, 5–12].

Many factors determines the differentiation of critical gap times and follow-up times values occurring at individual objects. One of the most important is the role of the road in the transport network and the nature of its surroundings, spatial development and functions of the area through which the road runs. These and many other factors affect above all the existing traffic volume, road equipment and shape of intersection, allowed speeds, driving comfort, relations, motivation and purpose of the journey and the way of driving [1, 4, 7, 8, 13–15]. Professional drivers who have to carry out transport tasks or commuters travelling to work or recreational vehicles behave in a different way [9, 13]. However, there is no information in the literature on the influence of this behavioural variation on the values of critical gap times and follow-up times at the analysed type of intersection.

In the paper an attempt was made to determine the critical gap times and follow-up time values at median, uncontrolled T-intersections with two two-lane pavements of the main road. The analyzed type of objects is not well known and described, all the more so there is no information about the values of gap time in the highest rank stream, which would allow vehicles from subordinate relations to join the traffic. Publicly available literature has little reference to this problem [4, 6, 7, 15].

The further part of the article presents and discusses the results of the study of critical gap times and follow-up times for all subordinate relations occurring at selected two intersections located in built-up areas and beyond built-up areas in the area of the Upper Silesian agglomeration. The necessary concepts were also collated and the factors influencing the values of critical gap times and follow-up times were mentioned.

2 Critical Gap Time and Factors Influencing the Value

Vehicle traffic can be considered equivalently in terms of time or distance, as shown in Fig. 1 showing the relationship between the gap and the headway in relation to the actual traffic situation at the uncontrolled intersection of a two-lane minor road with a two-lane one-way major road [2, 4, 7, 16–18]. Spatial dependencies have a subscript "L", whereas the temporal relations have the subscript "T". Length of car "LC" is a dimension that binds headway and gap.

The headway H_T shown in Fig. 1 is the time interval between passing the measuring station by a fixed element of the body of two consecutive vehicles. The gap G_T is measured as the time when, in the fixed measuring section, perpendicular to the axis of the major road, there is no bodywork element of two consecutive vehicles. The lag L_T defines the time remaining to be reached the measuring station by the nearest oncoming vehicle on the major road but in this case the measuring station is created on the superior road by the symmetry axis of the vehicle awaiting on the minor road approach. The follow-up time HQ_T is the time elapsing between the crossing of the stop line by a fixed body element of two consecutive vehicles waiting in the queue using the same gap in the highest priority stream to continue their journey [2, 6, 7, 16–19].

Drivers waiting at the minor road approach observe subsequent vehicles moving on the major road and the available gaps G_T between them. The gaps used by drivers to cross or merge in traffic in the major stream are called accepted, ones while others are rejected. According to the definition of Raff, the critical gap time (t_g) is the minimum time gap in the major stream accepted by the 50th quintile of the driver population waiting at minor-street approach. On the other hand, the follow-up time (t_f) is the time between crossing the stop line by a fixed element of the body of two consecutive vehicles using the same gap in the major stream to cross or merge the major stream in a situation where the constant queue at the minor street approach remains.

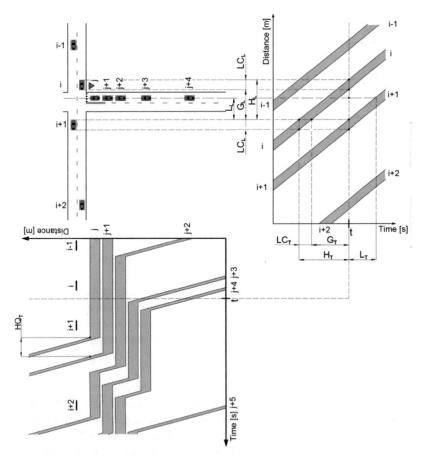

Fig. 1. Illustration of spatial and temporal relations in the analysis of the operation of the uncontrolled intersection to determine the critical gap time and the follow-up time

Noting the duration of gaps G_T: rejected and accepted by subsequent drivers in many situations of entering the traffic allows us to determine the value of the average gaps in the collected research sample. We must based on statistical analysis. That results characterize the behaviour of the driver population at a specific intersection or at the type of tested intersection, depending on the research extent. These determined values are t_g and t_f times, which characterise the population as a whole [2, 14, 16, 19, 20]. In the article for the specific objects, the values of times t_g and t_f were calculated separately for both examined objects and each subordinate relation based on observation of the vehicle sample and using the Siegloch method and the linear regression contained therein.

Nowadays, the recommended for calculating the values of t_g and t_f times have been varied only due to the relation, the manner of subordinating the approach and the nature of the surrounding on which the object is located. Additional simplified modifications are introduced to reflect the variation in visibility, the share of heavy vehicles and the longitudinal gradient of the minor road approach [1, 5, 16, 17, 19]. In fact, observed in practice values of critical gap times and follow-up times are influenced by a large set of factors strongly dependent on local and temporary conditions. Therefore, one should always strive to determine the possibly precise values of t_g and t_f times [1, 4, 14, 17, 18, 20]. Table 1 presents the division of these factors into permanent and variable.

Table 1. Identification and distribution of factors determining critical gap time and follow-up time values (Source: [1, 4, 7, 13, 15, 17, 21, 22]).

Mode of influence of factors	
Fixed	Variable
Vehicle type	Traffic conditions on major street
Nature of the traffic	Air transparency
Intersection geometry	Lighting
Number and interconnections of the approaches	Dynamic limitation of visibility (by parked vehicles and pedestrian traffic)
Sight	Psychomotor characteristics of the driver
	Time of day
	Pavement conditions
	Atmospheric conditions

The geometry of the intersection is a very broad issue that strongly determines the values of critical gap times and follow-up times. The number of lanes, traffic organisation, flared minor-street approaches, channelization, separate islands, run of the intersecting roads in the plan and in longitudinal profile influences this, the values and impact strength of overriding intensity and structure of traffic flows observed by waiting drivers. The number of higher-rank streams and their lane width determines the number, length and distribution of t_g and t_f times [4, 6–9, 14, 19, 20].

The number and angle of intersection of approach axes determines necessary for the observation: the number of superior relations and the angle of this observation, since in practice approaches intersect at angles other than 90°, which may increase or decrease visibility. The manoeuvres to enter such escapes do not have to be clearly to the right, left or straight ahead. In addition, the number of approaches determines the number of impeding, conflicting subordinate movements. The angles of longitudinal of the minor-street approaches are also important. Due to local conditions, the angles of intersection of traffic flows and the resulting relative speed are used for the description. A good position and pre-positioning of a subordinate vehicle to join the major-street stream give small angles of intersection. It causes less interference in traffic and acceptance of shorter gaps, at the same time influencing the severity of road events [3, 7, 9, 18, 21].

Sight in the vicinity of an intersection influences the behaviour of drivers awaiting and making decisions at the approach, as well as the way they enter the traffic and the resulting values of accepted gaps. This is both the value of the sight distance and the triangle of sight in the area of the intersection disc, the possibility of observing the traffic situation at the major-street approaches both by the driver of the first vehicle in queue and by the subsequent vehicles standing further away from the stopping line.

A variable factor influencing the way and the possibility of observing the traffic situation by drivers waiting at the minor-street approach is the lighting of the facility directly related to weather conditions, air transparency, cloudiness, the presence of large objects near the intersection casting shadow in a specific time. The impact of sunlight on awaiting at the minor-street approach drivers should also be emphasized (especially the possibility of dazzling).

Therefore, as presented in this chapter, the set of factors determining t_g and t_f times is huge and depends very strongly on local and momentary conditions. Therefore the estimation of values of critical gap time and follow-up time is always stochastic, based on statistical analysis, influencing on the compliance of the model with the observed real traffic conditions at the intersection. We should always strive to determine data as precisely as possible [1, 4, 17, 18, 20].

3   Median, Uncontrolled T-Intersection

Median, uncontrolled T-intersections with two two-laned one-directional pavements of major street being the subject of the research are characterized by a significant separateness from all other uncontrolled object types, which is manifested already in the analysis of the hierarchy of relations [4, 5, 7, 14, 20]. As it has been noted, these objects are not sufficiently characterized in the commonly available literature. Figure 2 shows the traffic organisation on this type of intersection. At the intersection type under study there are only 3 levels of movement hierarchy. Vehicles driving through and turning right from the major road are not expected to incur delay. All other streams are subordinate and it is possible to set t_g and t_f time for them. The right-hand turn from approach C is the same as this manoeuvre at intersections of the 1×2 type roads, with the conflicting flow rate includes only the stream going through from approach A on the right lane. Similarly, the relation marked as BL in the Fig. 2 can be considered classically, but drivers of this subordinate relation need to cross two lanes in the collision area and entails a longer evacuation time from the intersection, which must be taken into account by the drivers [4, 7, 17, 18].

The presence of a wide median strip and the influence of impedance make it possible to consider two-stage turn left from the minor road as two independent relations for the purpose of determining critical gap times and follow-up times, and this is also the most specific element that characterises a given type of intersection. Here, the first stage of the turn to the left is impedanted differently than at the four-way intersections with a wide median strip [14, 16–18]. Crossing the major road, which is the first part of the left turning marked CL1, in practice corresponds more closely to the through movement at a 1×2 intersection with only one highest priority stream need to be observed. Additionally this ones movement using two lanes. Equally imprecise is the adoption of the critical gap times and follow-up times recommended in the instructions for left turn manoeuvres when describing the second stage of a turn from a minor street. The drivers of the relation marked CL2 in the picture enter the traffic observing only the part of the stream going through from the major-street approach B appearing on the left lane, unlike in the HCM method, where the t_g and t_f times were set as for the left turning, in which the drivers have to observe the streams of as many as four higher priority movements [4, 5, 7].

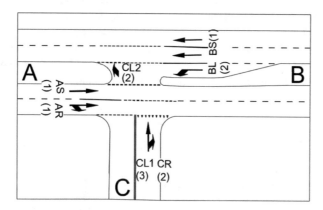

Fig. 2. Rank of streams at median, uncontrolled T-intersection (numbers in brackets shown rank of each movement)

Intersections of the examined type occur quite rarely in the road network and due to a large number of different structures of these objects and their location in many different areas, each of them is characterised by different specifics [3, 4, 9]. These objects are mainly found on high technical class roads and out of all geometric solutions, they remain the most clear, understandable, safest, adapted to the characteristics of the roads that form it. They are always shaped according to local needs and in a minimal way (due to channelization and median strip), while maintaining legibility. They lead traffic in rank 1 relations much smoother than in the case of the use of traffic lights, while at the same time making it easier for subordinating relation vehicles to enter the traffic by means of channelization [3, 4, 9]. Traffic efficiency and increased safety have been achieved through the use of a wide median strip, the distribution of through movement on the major road on two lanes, a reduction in the impact of some streams through channelization, improved visibility by reducing distraction and blinding drivers by vehicles travelling in opposite direction. In addition, it affects the stopping points and trajectory of the manoeuvres, which allows for unambiguous location of collision points in the space, reduction of their area and shortening the time of vehicles passing through the conflict zone and proper creation of angles of trajectory intersection. Limiting the scope of the track reduces the number of decisions to be taken [3, 4, 7, 9].

4   Characteristics of Selected Objects and Conducted Research

Figures 3 and 4 presents a diagrams of the tested intersections together with the organization of traffic including pavements marking and signs. The subject of the analysis were two intersections located in the Upper Silesian agglomeration. Research object no. 1 is located in one of the central districts of the voivodship city of Katowice, densely built-up, performing commercial and service functions. The major road is an important link between the northern and southern districts of the city and the city centre. It has a meridian mileage between the A4 motorway and the Drogowa Trasa Średnicowa express road, which additionally emphasizes its importance not only local, inter-district, but also in the scale of the whole agglomeration.

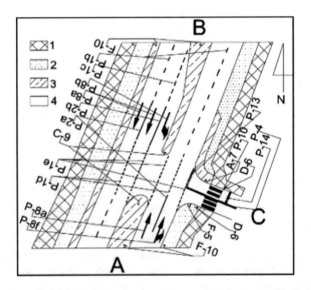

Fig. 3. Diagrams of the tested intersection: in built-up area (object no. 1) (Surface markings: 1 - sidewalk, 2 - grass, 3 - concrete paver, 4 - pavement)

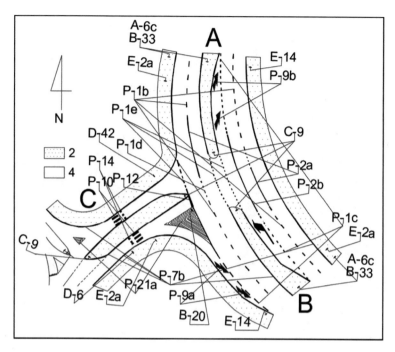

Fig. 4. Diagrams of the tested intersection outside the built-up area (object no. 2) (Surface markings: 2 - grass, 4 - pavement)

The described location of the object in the built-up area is presented in Fig. 5 therefore outside the built-up area in Fig. 6.

Fig. 5. Location of research object no. 1 in the transport system of a selected part of the Katowice city (Source: [23])

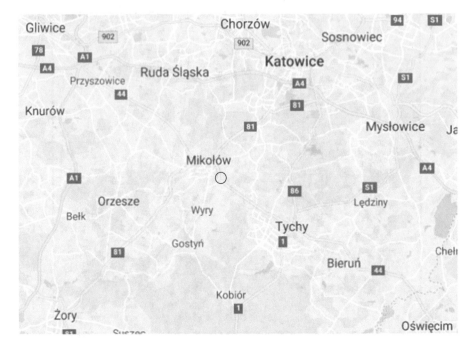

Fig. 6. Location of research object no. 2 in the transport system of the region (Source: [23])

Research object no. 2 is located at the edge of Mikołów (Śląskie Voivodeship), on the transport route between such important urban centres in the region as Gliwice, Katowice and Tychy. The intersection is located at the intersection of the road connecting Gliwice and Tychy and the road running southwards through the whole southern part of the city and connecting with two express roads significant in the scale of the region. This indicates the transit character of the distinguished area, characterized by a significant size of traffic flows in many relations, which is additionally evidenced by the course, technical class and the number of roads. The neighbourhood of large agglomerations constituting residential, manufacturing and commercial-service centres results in the generation of trips in many motivations and directions. It is an intersection of high regional and supra-regional importance. The described location of the research object no. 2 in the region is presented in Fig. 5.

The basic similarities of the two objects include the geometric system consisting of two two-lane one-directing pavements at major road and a short auxiliary lane to the left turn from the major road and a two-lane pavement of the minor road with a flare for right-turning movement. Both objects, due to their role, are clearly separated from the surroundings. There is no pedestrian impact. The major road as well as the minor road on both objects run horizontally in the profile. In the plan, however, only outside the built-up area there is an arch on the major road within the intersection, however, with a significant radius. The impact of reduced sight of the traffic situation in each relations is comparable between the two objects. In built-up area, it results from running roads in a cutting, the presence of buildings and railway embankment in the vicinity, while outside the built-up area, it results from a horizontal arch on the major road.

The differences result primarily from the location, the function performed in the road network, the impact and role of the surrounding and the traffic to which their geometry has been adapted.

Table 2 presents features differentiating the analyzed objects.

Table 2. List of differences characterizing the studied objects.

Feature	Research object no. 1	Research object no. 2
Location	In the built-up area	Outsider the built-up area in the vicinity of agglomeration
Movement to be handled	Internal, external source and target traffic	External source and target, transit traffic
Surrounding	Dense buildings, low, multi-storey tenement houses. Mixed, housing, industrial, manufacturing, service and commercial functions	Low, multi-storey tenement houses. Mixed, industrial, manufacturing, service and commercial functions. Green and agricultural areas
Acceleration lane	Lack	Occurs
Layout in the plan	Intersection of the roads at an angle 90°	The object is located in the middle of the road arch. Inlet of a subordinate straight path on the outside of the curve
Sight	Limited observation of the major road by successive vehicles at the minor-road approach. On the stopping line, difficulties related to the bridgehead of the flyover and traffic barriers and obstruction by vehicles placed in parallel	Limited by the break of the major road in the plan and profile

Figure 7 presents a general scheme of procedures for the determination of critical gap time and follow-up time according to the Siegloch method.

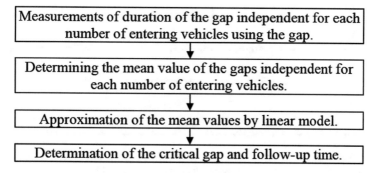

Fig. 7. Siegloch's method - best practice

It is therefore clear that the process has four stages. Firstly it must be registered a statistically significant number of vehicles of subordinate streams entering the traffic under initial conditions defined in the assumptions of the Siegloch method. Secondly the duration of gaps G_T: rejected and accepted as well as the number of vehicles using should be determined separately for each relation. Then, the data prepared in such a way together with the determined average values of these times are the basis for drawing up the graph, examples of which for research objects are presented in Figs. 8 and 9.

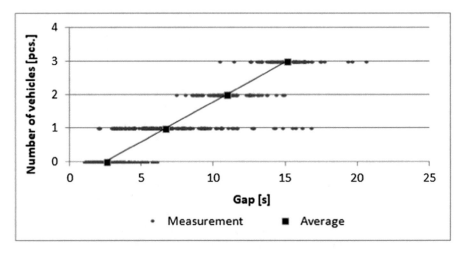

Fig. 8. Graphical presentation of collected measurement data and linear regression model for right-turning from minor-road approach relation of research object no. 1

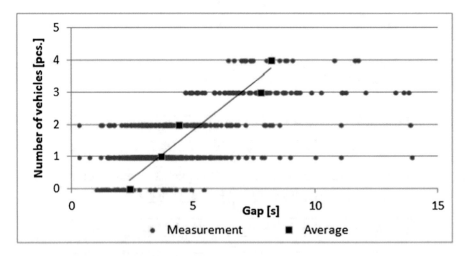

Fig. 9. Graphical presentation of collected measurement data and linear regression model for right-turning from minor-road approach relation of research object no. 2

The research was conducted in spring 2018, on Wednesdays and Thursdays in sunny weather. The existing immediate surroundings of the objects were used to observations from a height of about 6 metres above the roadway level. The object was monitored with the use of cameras. The collected film material, including 16 h (4 measurement days) per each object, was analyzed. All movement situations consistent with the assumptions of Siegloch's method were used, what individualized the results for local objects.

The use of a linear regression model for the average duration of gaps used by 0, 1, 2 up to n vehicles respectively, entering or crossing the major stream, allows the determination of the approximation equation. "N" means the maximum registered number of vehicles of a given subordinate relation which have used the same gap in the highest priority movement to enter traffic. The coefficients of the linear regression model contains information about the critical gap time t_g and follow-up time t_f, which generally characterize the entire population of drivers.

In addition, the comparison of both diagrams in Figs. 8 and 9 shows the differences in the behaviour of drivers of the right-turn movement from the minor-street approach and the differences in the test sample collected in both cases. You can see that the drivers at object 2 use the shorter gaps in the major-street stream to continue driving.

5   Test Results

Critical gap times and follow-up times were determined for all four subordinate relations occurring at the crossroads under study. The results presented in Table 3 characterize local objects.

Table 3. Values of critical gap time and follow-up time [s] for all subordinate relations.

Parameter		Research object no. 1				Research object no. 2			
		BL	CR	CL1	CL2	BL	CR	CL1	CL2
Follow-up time	t_f	3.79	4.19	4.97	3.79	2.40	1.56	3.52	2.96
Critical gap time	t_g	3.78	4.68	5.88	2.38	4.22	2.94	4.94	6.15

The analysis of the data presented in the Table 3 clearly indicates that the considered intersections differ from the most popular objects located at the intersection of roads in the 1 × 2 system. The differences in values and tendencies between the two studied objects result primarily from the nature of the traffic they handle and the different behaviours of individual drivers. The first, most general, noteworthy observation is that for all subordinate relations, the follow-up times t_f are at a intersection located outside built-up areas smaller than in the city centre. This is due to the greater fluidity of traffic on transit roads, higher permitted speeds and the maintenance of longer distances between vehicles in the highest priority movement and the way in which they are grouped.

At both intersections, the critical time for the individual relations are greater than the follow-up times. However, in built-up area the variation in t_g times compared to t_f is smaller than outside built-up area, i.e. each subsequent vehicle entering from the queue individually considers the possibility of continuing driving and needs more time to make a decision differently than outside built-up areas, where shorter entry intervals for subsequent vehicles have been noted.

A detailed analysis of the respective relations on the two examined objects indicates that in built-up area drivers are more likely to risk, more often break the rules and enforce priority, while outside built-up area are more restrained, carefully selecting and using the available gaps to continue driving. The presented trends are best visible for the relation BL, for which the values t_g and t_f are similar to each other in built-up area, with t_g being smaller and t_f bigger than those corresponding to them outside the built-up area. The influence of the surrounding and the road function is primarily determined here by the assumed geometry of the traffic corridor for vehicles of a given relation. This phenomenon is also visible for the CL1 relationship, where the width of the lanes, speed of cars and the trajectory of vehicle traffic meant that there is a significant difficulty in joining to traffic in built-up area. Therefore, higher values assume critical gap time and follow-up time at object no. 1 in comparison with object no. 2. The relation CL2 is different from the presented above. This results directly from the features of the type of intersection considered and an independent analysis of this turn to the left. The lower value of t_g time in built-up areas is related to the already described drivers' propensity for risk and more aggressive driving. This is also influenced by smaller differences in the speed possessed by the two streams. However, the greater value of time t_f is determined by the caution of successive drivers entering the queue of evaluators, regardless of the previous vehicles, the opportunity to join the traffic.

The last observation made is the demonstration of the significant impact of the acceleration lanes on reducing the t_g and t_f times. In the examined example, the application of this solution allowed to obtain a decrease in their value by about 50% due to the decrease in the relative difference in speed that the interacting streams have. As a result, drivers who are entering into the major road stream often use shorter time gaps.

6   Conclusions

The article presents a comprehensive collection of factors influencing the values of critical gap times and follow-up times. They have been discussed in a theoretical manner and selected ones occurring on the examined real objects, i.e. nature and function of the surrounding identified and assessed due to the impact on the efficiency of traffic. The conducted field studies and literature analysis clearly showed the differentness of median, uncontrolled T-intersections dividing them from objects created by crossing the 1×2 type roads. Analysis of the existing relations ordered to separate the examined intersections from objects of the same type but with four approaches.

The research results indicated that the nature and functions of the surrounding directly affect the values of the critical gap times and follow-up times by determining the width of the lanes, the designed trajectories of vehicle and the permitted speeds. It was also pointed out that outside the built-up area, the traffic is more fluid, as subsequent drivers joining the traffic in the same gap need smaller time intervals t_f for manoeuvring than in built-up area. In the city drivers drive aggressively, more often breaks regulations and enforce priority, using shorter time gaps to join traffic and need more time to decide on the use of the same gap by the next drivers. The availability of an acceleration lanes means that a shorter time is sufficient for subordinate vehicles to continue their journey.

However, it should be emphasized that the nature of surrounding is not the only factor differentiating the examined objects. However, it has an important role at these intersections. Taking into account the available small testing ground, it is possible to draw the right conclusions regarding the impact of surrounding character on the values of the critical gap times and follow-up times. However, in order to obtain results characterizing a given type of T-intersection, it is necessary in general to conduct further in-depth tests on a larger number of objects.

However, the obtained test results showed that t_g and t_f times determined individually for a local object or characterizing the selected object type can be used as an effective tool for calibrating the capacity calculation method and assessment of traffic conditions and this will contribute to better spatial planning and traffic management.

References

1. Abhigna, D., Kondreddy, S., Shankar, K.: Effect of vehicle composition and delay on roundabout capacity under mixed traffic conditions. Arch. Transport **40**(4), 7–14 (2016)
2. Burnos, P., Gajda, J., Sroka, R.: Pomiary parametrów ruchu drogowego. Wydawnictwo PWN, Warszawa (2015)
3. Hobbs, F., Richardson, B.: Traffic engineering, vol. 2. Pergamon Press, London (1967)
4. Gaca, S., Suchorzewski, W., Tracz, M.: Inżynieria ruchu drogowego. Teoria i praktyka. Wydawnictwo Komunikacji i Łączności, Warszawa (2011)
5. Transportation Research Board: Highway Capacity Manual. National Research Council. Transportation Research Board, Washington (2000)
6. Akcelik, R.: A Review of Gap-Acceptance Capacity Models. University of South Australia, Adelaide (2007)
7. Gerlough, D., Huber, M.J.: Traffic Flow Theory. Transportation Research Board. National Research Council. Transportation Research Board, Washington (1975)
8. Ramu, A., Hari, K.G., Lakshimi, D., Rao, K.R.: Comparative evaluation of roundabout capacities under heterogeneous traffic conditions. J. Modern Transp. **23**(4), 310–324 (2015)
9. Bunden, W.: Introduction to Traffic Science. Printerhall Ltd., London (1967)
10. Macioszek, E.: Analysis of significance of differences between psychotechnical parameters for drivers at the entries to one-lane and turbo roundabouts in Poland. In: Sierpiński, G. (ed.) Intelligent Transport Systems and Travel Behaviour. AISC, vol. 505, pp. 149–161. Springer, Cham (2017)
11. Macioszek, E.: The comparison of models for critical headways estimation at roundabouts. In: Macioszek, E., Sierpiński, G. (eds.) Contemporary Challenges of Transport Systems and Traffic Engineering. LNNS, vol. 2, pp. 205–219. Springer, Cham (2017)

12. Macioszek, E.: The comparison of models for follow-up headway at roundabouts. In: Macioszek, E., Sierpiński, G. (eds.) Recent Advances in Traffic Engineering for Transport Networks and Systems. LNNS, vol. 21, pp. 16–26. Springer, Cham (2018)

13. Hobbs, F., Richardson, B.: Traffic Engineering, vol. 1. Pergamon Press, London (1967)

14. Thamizh, A., Reebu, Z.K.: Methodology for modelling highly heterogeneous traffic flow. J. Transp. Eng. **131**, 544–551 (2005)

15. Devarasetty, P., Zhang, Y., Fitzpatrick, K.: Differentiating between left-turn gap and lag acceptance at unsignalized intersections as a function of the site characteristics. J. Transp. Eng. **138**, 580–588 (2011)

16. Mohan, M., Satish, Ch.: Review and assessment of techniques for estimating critical gap at two-way stop-controlled intersections. European Transport **61**(8), 1–18 (2016)

17. Manish, D., Mokkades, A.: Gap acceptance behavior of drivers at uncontrolled T-intersections under mixed traffic conditions. J. Modern Transp. **26**(2), 119–122 (2017)

18. Ashalatha, R., Chandra, S.: Critical gap through clearing behavior of drivers at unsignalised intersections. J. Civil Eng. **15**(8), 1427–1434 (2016)

19. Brillon, W., Koenig, R., Troutbeck, R.: Useful estimation procedures for critical gaps. Trans. Res. Part A: Pol. Pract. **33**(3), 161–186 (1999)

20. Vasconcelos, A.L., Seco, A., Silva, A.: Comparison of procedures to estimate critical headways at roundabouts. Promet-Traffic Transp. **25**(1), 43–53 (2013)

21. Yan, X., Radwan, E.: Influence of restricted sight distances on permitted left-turn operation at signalized intersections. J. Transp. Eng. **134**, 68–76 (2008)

22. Zohdy, I., Sadek, S., Rakha, H.: Empirical analysis of effects of wait time and rain intensity on driver left-turn gap acceptance behavior. Transp. Res. Rec. J. Transp. Res. Board **2173**, 1–10 (2010)

23. Open Street Map. https://www.openstreetmap.org/

Long-Distance Railway Passenger Transport in Prague - Dresden/Cheb Relations

Rudolf Vávra$^{(\boxtimes)}$ and Vít Janoš

Faculty of Transportation Sciences, Czech Technical University in Prague,
Prague, Czech Republic
{vavrarud, janos}@fd.cvut.cz

Abstract. This paper is focused on testing of technological possibilities of variants of transport concept of long-distance railway passenger between Praha - Ústí and Labem - Dresden/Cheb. The aim is to check if it makes sense to strengthen transport supply in these relations even on current railway infrastructure. At first, there is described the existing operational conception in researched relations from the point of view of technical and technological boundary conditions. Then, there are analyzed these relations in the area of transport demand, on which basis the extent of supply is tested for new conceptual possibilities of these long-distance lines. Formulation of technological conditions restricting construction of trains paths of these lines follows, whereupon the individual variants are checked from the constructional point of view. Designed model periodic paths are subsequently assessed from interactions and technological conflicts viewpoints, which results to recommendation of the most operationally appropriate solution.

Keywords: Long-distance railway · Passenger transport · Transport demand · Transport supply · Periodic timetable · Train paths · Timetabling · Rail capacity

1 Introduction

Attractivity of railway connections (generally of public transport connections) is affected not only by travel time but also by frequency of these connections and by number of transfers. The frequency of (ideally direct) connections in the long-distance railway relations which are justified by transportation relations and by travel times, more precisely by their comparison with road transport (whether individual car or public bus transport), should be standardly at least in 60-min period (at least during peak hours).

This paper is focused on long-distance passenger transport on lines Praha - Ústí nad Labem - Dresden (and then Berlin/Hamburg) and Praha - Ústí nad Labem - Cheb. These two lines currently operate in 120-min interval and they are mutually interposed in Praha - Ústí nad Labem section into common 60-min interval. For solving mutual interactions of these lines, there is necessary to consider following railway lines:

© Springer Nature Switzerland AG 2020
E. Macioszek and G. Sierpiński (Eds.): Modern Traffic Engineering in the System Approach to the Development of Traffic Networks, AISC 1083, pp. 257–274, 2020.
https://doi.org/10.1007/978-3-030-34069-8_20

- 090 Praha - Ústí nad Labem - Děčín,
- 130 Ústí nad Labem - Klášterec nad Ohří,
- 140 Klášterec nad Ohří - Karlovy Vary - Cheb,
- Děčín - Schöna - Bad Schandau - Pirna - Dresden.

On these railway lines there currently operate in 120-min interval these long-distance lines:

- EuroCity trains of line Ex3 Praha - Ústí nad Labem - Děčín - Germany,
- fast trains of line R5 Praha - Ústí nad Labem - Karlovy Vary - Cheb,
- fast trains of line R20 Praha - Ústí nad Labem - Děčín (during peak hours in peak direction there is one-hour-interval).

In following chapters current operational conceptions of these lines are initially described and the analysis of origin-destination pairs is accomplished, which are supposed to justify operation of these lines (at least during peak hours, at least in some sections of their route) in 60-min interval even on current railway infrastructure. It is followed by formulation of technological conditions restricting the construction of train paths of the researched long-distance lines. Then the paths of these lines are proposed (on current railway infrastructure) according to the technological possibilities in different variants (achievement of different IPT-nodes, using of different train sets etc.). Afterwards, the paths construction is evaluated and then there are created variants of operational concepts from suitable paths in order to lay out the transport supply in time so that as many connections as possible could profit from it (the evener layout of the supply in time is used, the bigger profit for served relations exists).

2   Analysis

In the analytical chapter there is initially described the current operational concept of the considered lines, afterwards the transportation relations in the researched area are characterized.

2.1   Description of Current Operational Conception of the Considered Long-Distance Lines

Line Ex3 Praha - Ústí nad Labem - Děčín - Germany. The north branch of line Ex3 Praha - Ústí nad Labem - Děčín - Dresden - Berlin/Hamburg/Kiel/Leipzig (it will be renamed as Ex5 since timetable 2019/20 comes into force) performs function of the first/highest (express) transportation segment - connections of this line stop between Praha hl.n. and Dresden Hbf at stations: Praha-Holešovice, Ústí nad Labem hl.n., Děčín hl.n. and Bad Schandau [1, 2].

The line is operated in 120-min interval and it is constructed to be interposed with line R5 Praha - Ústí nad Labem - Cheb into common 60-min interval between Praha hl. n. - Ústí nad Labem hl.n.. Line Ex3 reaches node Praha hl.n. at minute 30 (departure at E:30/arrival at O:30; E = even hour, O = odd hour), in Ústí nad Labem it occurs at time position O:45 (in direction to Dresden)/E:15 (in direction to Praha), in Dresden it reaches the IPT-node at O:00 [2].

Line R5 Praha - Ústí nad Labem - Cheb. Line R5 Praha - Ústí nad Labem - Cheb (it will be renamed as line R15 since timetable 2019/20 comes into force) performs function of mixed transportation segment: between Praha hl.n. - Ústí nad Labem hl.n. it performs function of the first (express) transportation segment, between Ústí nad Labem hl.n. - Karlovy Vary - Cheb it performs function of the second transportation segment (with more stops). Connections of this line stop between Praha hl.n. and Cheb at stations and stops: Praha-Holešovice, Ústí nad Labem hl.n., Teplice v Čechách, Bílina, Most, Jirkov zastávka, Chomutov město, Chomutov, Kadaň-Prunéřov, Klášterec nad Ohří, Ostrov nad Ohří, Karlovy Vary, Chodov, Sokolov and Kynšperk nad Ohří [1, 2].

The line is operated in 120-min interval. In Praha hl.n. - Ústí nad Labem hl.n. section it is interposed with line Ex3 into common 60-min interval. It reaches node Praha hl.n. at minute 30 (departure at O:30/arrival at E:30), station Ústí nad Labem hl. n. is reached at time position E:45 (in direction to Cheb)/O:15 (in direction to Prague), station Most is reached at minute 30 (O:30 in direction to Cheb/E:30 in direction to Prague), station Kadaň-Prunéřov at time position E:00, station Karlovy Vary at time position E:45 (in direction to Cheb)/O:15 (in direction to Prague) and node Cheb is reached at minute 30 (arrival at O:30, departure at E:30) [2].

Line R20 Praha - Ústí nad Labem - Děčín. Line R20 Praha - Ústí nad Labem - Děčín performs function of the second transportation segment. Connections of this line stop between Praha hl.n. and Děčín hl.n. at stations and stops: Praha-Holešovice, Praha-Podbaba, Kralupy nad Vltavou, Hněvice, Roudnice nad Labem, Bohušovice nad Ohří, Lovosice and Ústí nad Labem hl.n. [1, 2].

The line is operated basically in 120-min interval, but during peak hours there is offered 60-min interval in peak direction (in the morning to Prague, in the afternoon from Prague). Path construction of the line is determined by reaching IPT-node Lovosice at E:00, which means reaching of station Praha hl.n. at time position E:50 (departure)/O:10 (arrival), station Ústí nad Labem hl.n. at time position E:15 (in direction to Děčín)/O:45 (in direction to Prague) and station Děčín hl.n. not sharply at minute 30 (arrival after E:30/departure before O:30) [2].

2.2   Characteristic of Transportation Relations of People

The towns which are connected by long-distance lines Ex3, R5 and R20 were chosen for analysis of transportation relations (each stop of these lines was matched with a municipality, on which territory the stop is located, or with additional bigger town, which is with this municipality related from geographical and transportation viewpoints). These municipalities were chosen for the analysis of transportation relations: Praha, Kralupy nad Vltavou, Štětí, Roudnice nad Labem, Bohušovice nad Ohří, Terezín, Lovosice, Litoměřice, Ústí nad Labem, Děčín, Dresden, Teplice, Duchcov, Bílina, Most, Litvínov, Jirkov, Chomutov, Kadaň, Klášterec nad Ohří, Ostrov, Karlovy Vary, Chodov, Sokolov, Kynšperk nad Ohří, Cheb. In case of line Ex3 Berlin and Hamburg were chosen additionally beyond the researched section (Praha - Dresden).

Domestic Transportation Relations. For analysis of domestic transportation relations publication *Population and Housing Census* (in Czech *Sčítání lidu, domů a bytů*) [3] from year 2011 was used. It is necessary to mention disadvantages of using of this publication for analysis of transportation relations:

- only commuting to work and schools are mentioned, without other transportation relations,
- because of questionnaire way of getting the data, there are possibly included incorrectly filled census forms too, which is then reflected in output data,
- the data ale old (from year 2011), newer ones are not available (the following census is going to be made in year 2021), thus the data do not reflect current state.

Table 1 displays strong long-distance transportation relations (with number of commuting people higher than 500) which result from the analysis of domestic transportation relations in the researched area. This table also displays a comparison of travel times in these relations when using train (lines Ex3, R5 or R20), when using a car (according to a route planner) and a bus (using of direct connections or in case of combination of bus and Prague urban transport using of connections with one transfer). These travel times were used for calculation of coefficients of travel times which are calculated as the travel time by train divided by the travel time by car or by bus (coefficients of travel times lower than 1.25 have green background color).

Table 1. Long-distance domestic relations in the researched area with more than 500 commuting according to the *Population and Housing Census* [3] and comparison of travel times in these relations when using railway transport (lines Ex3, R5, R20), car and bus public transport [2, 4, 5]. Coefficients of travel times lower than 1.25 have green background color.

Relation	Commuting according to [3]	Travel time by train (min)	Travel time by car (min)	Coefficient of travel time train/car	Travel time by bus (min)	Coefficient of travel time train/bus
Děčín - Praha	587	85.5	80.5	1.062	–	–
Teplice - Praha	640	91.5	58.0	1.578	73.5	1.245
Karlovy Vary - Praha	648	188.5	101.5	1.857	132.6	1.422
Chomutov - Praha	679	135.5	77.0	1.760	118.0	1.148
Roudnice n. L. - Praha	702	53.5	35.5	1.507	46.7	1.146
Litoměřice - Praha	764	84.5	47.5	1.779	75.2	1.124
Most - Praha	954	116.5	71.5	1.629	102.6	1.135
Ústí n. L. - Praha	1 246	68.3	57.5	1.188	71.3	0.958

Transportation Relations between the Czech Republic and Germany. For the analysis of the cross-border transportation relations *Population and Housing Census* [3] publication is not convenient. For this purpose, *Transtools* [6] was used, which is a transportation model of Europe from year 2005. From analysis of transportation relations viewpoint *Transtools* have these shortcomings:

- it is only a model [7], differently high differences from real state can exist (however, the model is supposed to be sufficiently calibrated),
- it is a model from year 2005, but the system is still being progressed, therefore the numbers of passengers do not have to reflect actual state.

It is important to mention that values gained from the *Transtools* model are not comparable with the values gained from the *Census of people, houses and flats* publication, because the second ones include only people commuting to work and school, but the first ones include all types of transportation relations.

Table 2 displays transportation relations between Prague and the German cities Dresden, Berlin and Hamburg according to the *Transtools* model (sum of both directions and of all transport modes - railway, road and air transport). Then this table displays similar comparison of travel times as was made above at domestic transportation relations.

Table 2. Selected long-distance cross-board relations between the Czech Republic and Germany and comparison of travel times in these relations when using railway transport (lines Ex3, R5, R20), car and bus public transport [2, 4–6]. Coefficients of travel times lower than 1.25 have green background color.

Relation	Transportation demand according to Transtools 2005	Travel time by train (min)	Travel time by car (min)	Coefficient of travel time train/car	Travel time by bus (min)	Coefficient of travel time train/bus
Praha - Dresden	2 910	136.0	88.0	1.545	115.7	1.175
Praha - Berlin	12 403	247.0	204.0	1.211	273.2	0.904
Praha - Hamburg	5 873	396.0	354.5	1.117	595.6	0.665

Assessment of the Analysis of Transportation Relations. The analysis of transportation relations and the comparison of travel times show that the lines should be operated in 60-min interval (at least during peak hours) [8]:

- line Ex3 at least in Praha - Děčín section, in case of interest of Germany (either rail carrier DB or German Federal Ministry of Transport and Digital Infrastructure) it would be appropriate to implement 60-min interval even to Berlin (with respect to prospective shortening of travel time Dresden - Berlin),
- line R5 particularly between Praha - Ústí nad Labem - Chomutov.

It is also possible to consider additional extending of 60-min interval of line R20 at least between Praha - Ústí nad Labem.

3 Technological Conditions Restricting Train Paths Construction

In this chapter the restricting technological conditions for construction of train paths of lines Ex3, R5 and R20 are determined.

3.1 Periodic Timetable

Periodic timetable is a type of timetable with fixed interval between individual connections of all lines in the network. The interval could be also called as the period. The periodic (or integrated periodic) timetable (IPT) is characterized in comparison with interval timetable by united symmetry axis of all lines (in Central Europe the symmetry axis is commonly ca. at minute 00). The symmetry axis is the time, when the connections of opposite directions of a single line systematically meet each other [9].

An IPT-node is that place in the transport network where connections of multiple lines systematically meet each other. A main IPT-node and a secondary IPT-node can be distinguished. The main IPT-node is a such IPT-node where the connections of multiple lines in both directions meet each other (at the time of symmetry). The secondary IPT-node is an IPT-node where connections of multiple lines only in one direction meet each other (exactly between neighboring symmetry axes; in the second direction the connections meet each other one period later).

Lines Ex3, R5 and R20 Specific Requirements. Specific requirements during train paths construction of lines Ex3, R5 and R20, which are based on periodic timetable principles, are:

- with respect to the results of the analysis of transportation relations the train paths of the long-distance lines must be able to be constructed in 60-min interval,
- all the lines must have united symmetry axis ca. at minute 00 as common in Central Europe, more precisely the axis of symmetry will be based on Dresden S-Bahn, whose train paths will be respected as a fixed input condition,
- line Ex3 will reach IPT-node Dresden - at two variants: at X:00, more precisely at O:00 (as in current state), or at X:30 (in agreement with a plan of German Federal Ministry of Transport and Digital Infrastructure [10] for year 2030),
- line R5 will reach IPT-node Cheb - at two variants: at X:30 (as in current state), or at X:00, more precisely at O:00 (in agreement with a plan of German Federal Ministry of Transport and Digital Infrastructure [10] for year 2030),
- line R20 will reach IPT-node Lovosice at X:00, more precisely at E:00.

3.2   Other Lines Operated in the Researched Area

Construction of the train paths of lines Ex3, R5 and R20 is restricted by the other lines operated on the researched railway lines too. Following lines (and their time positions) restrict the searched solution.

Line R15 Ústí nad Labem - Děčín - Česká Lípa - Liberec. Long-distance line R15 restricts construction of lines Ex3 and R20. The 120-min interval of line R15 will be kept and its crossing at Benešov nad Ploučnicí station as well, which implies to keeping the line at about the same time position on railway line 090 as in current state. Keeping of the transfer binding between R5 and R15 at Ústí nad Labem hl.n. station will be ideally pursued in the variant in which line R5 reaches IPT-node Cheb at X:30.

Line R25 Plzeň - Žatec - Chomutov - Most. Long-distance line R25 restricts construction of line R5. The 120-min interval, IPT-node Most at O:00 and crossing at Žatec západ station will be kept at line R25, which implies to keeping the line at about the same time position on railway line 130 as in current state.

Line S1 of S-Bahn Dresden: Meißen - Coswig - Dresden - Pirna - Bad Schandau - Schöna. Line S1 of S-Bahn Dresden restricts line Ex3 in Pirna – Schöna section (in Dresden – Pirna section line S1 uses another double-track railway line than line Ex3). During construction of train paths of line Ex3 frequency and time position of line S1 connections will not be changed in any way, they will be taken as a fixed input restricting condition.

Line U1 Děčín - Ústí nad Labem - Most - Chomutov - Kadaň-Prunéřov. Line U1 restricts construction of all the researched long-distance lines. Constructional time position of line U1, which is based on reaching IPT-node Děčín at X:30, IPT-node Ústí nad Labem at X:00 and IPT-node Most at X:00, will be kept as well as its 60-min interval in Děčín hl.n. - Most section and 120-min interval in Most - Kadaň-Prunéřov section. At Děčín hl.n. station (at X:30) and at Most station (at E:00) short turns of the train sets will be kept as well (the minimal time for turning is 6 min). Not overtaking of line U1 in Děčín hl.n. - Ústí nad Labem hl.n. section and in Ústí nad Labem hl.n. - Most section by the researched long-distance lines is also desired, because travel times of relatively strong regional transportation flows in metropolitan region of Ústí nad Labem would be lengthen by the overtaking.

Line U2 Most - Chomutov - Klášterec nad Ohří - Karlovy Vary. Line U2 restricts construction of train paths of line R5. 120-min interval with reaching IPT-node Most at E:00, as well as transfer binding to line R25 at Chomutov station (for journeys from Klášterec nad Ohří to Žatec and) will be kept at the connections Most - Chomutov - Klášterec nad Ohří, which operation is based on periodic timetable (at least approximately). The 120-min interval of short Kadaň-Prunéřov - Klášterec nad Ohří connections with their short transfer bindings to line U1 at Kadaň-Prunéřov station will be kept too. Other connections of line U2 are operated unsystematically, therefore their train paths are not considered as restricting ones.

Line S4/U4 Praha - Kralupy nad Vltavou - Lovosice - Ústí nad Labem. Line S4/U4 restricts construction of all the researched lines. 60-min interval in Ústí nad Labem - Kralupy nad Vltavou section and 30-min interval in Kralupy nad Vltavou - Praha Masarykovo nádraží section, reaching of IPT-node Ústí nad Labem hl.n. at X:00 (including short turn of train set - minimal turning time is 6 min) and IPT-node

Lovosice at X:30 will be kept at this line. In Lovosice - Praha Masarykovo nádraží section the current time position will not be required, but it is desired to keep the line undivided by too long dwell times at intermediate stations and stops in whole Ústí nad Labem - Praha section. Not overtaking of line S4/U4 in Ústí nad Labem - Lovosice and Kralupy nad Vltavou - Praha sections by the researched long-distance lines is required as well, because travel times of relatively strong regional transportation flows in these relations would be lengthen by the overtaking.

Line U24 Ústí nad Labem - Teplice v Čechách - Litvínov. Line U24 restricts construction of line R5. In agreement to its prospective operational concept 60-min interval and reaching of IPT-node Ústí nad Labem hl.n. at time position X:30 (including short turn of train set - minimal turning time is 6 min) is required. It is also necessary that the U24 connections of opposite directions pass each other on the double-track railway line 130 (the limit way of passing is crossing at Oldřichov u Duchcova station).

Line U28 Děčín - Bad Schandau - Sebnitz - Dolní Poustevna - Rumburk. Line U28 restricts construction of train paths of line Ex3. 120-min interval, transfer binding to line S1 of S-Bahn Dresden at Bad Schandau station (at time position O:15/E:45), transfer binding to line U1 at Děčín hl.n. station and ideally transfer binding to a long-distance line in direction Ústí nad Labem - Praha will be kept at the U28 connections, which are operated in whole Děčín - Rumburk section. Short peak-hours connections Děčín - Schöna are not considered to be restricting the construction of line Ex3.

4   Proposal of Train Paths Variants

In this chapter there is described the train paths construction of the researched long-distance lines Ex3, R5 and R20 in possible variants. For timetabling software FBS (Fahrplanbearbeitungssystem) was used. The train paths are constructed for current state of railway infrastructure, no infrastructural modifications are supposed compared to the current state.

4.1   Line Ex3 Praha - Ústí nad Labem - Dresden

Train paths of line Ex3 were constructed for Siemens Vectron electric engine and 9 carriages of classic construction. Three variants of train paths were constructed in total, which are in terms of this paper named: DR00-A, DR00-B and DR30.

Train Path DR00-A. Train path DR00-A of line Ex3 reaches IPT-node Dresden Hbf at X:00. In direction from Prague to Dresden the construction of the path follows these points (in the opposite direction it is constructed symmetrically):

- pass through Lovosice station at the latest minimum departure headway before departure of a train of line U4 in direction Ústí nad Labem,
- pass trough Pirna station at the earliest minimum arrival headway after arrival of a train of S-Bahn Dresden S1 line from direction Bad Schandau.

Train Path DR00-B. Train path DR00-B of line Ex3 reaches IPT-node Dresden Hbf at X:00. In direction from Prague to Dresden the construction of the path follows these points (in the opposite direction it is constructed symmetrically):

- arrival at Ústí nad Labem hl.n. station at the earliest minimum arrival headway after arrival of a train of line U4 from Lovosice,
- departure at Ústí nad Labem hl.n. station at latest minimum departure headway before departure of a train of line U1 in direction to Děčín.

Train Path DR30. Train path DR30 of line Ex3 reaches IPT-node Dresden Hbf at X:30. In direction from Prague to Dresden the construction of the path follows these points (in the opposite direction it is constructed symmetrically):

- departure at Děčín hl.n. station at the latest minimum departure headway before departure of a train of line U28 in direction Bad Schandau - Rumburk,
- arrival in Pirna station at the earliest minimum arrival headway after arrival of a train of S-Bahn Dresden S1 line from direction Schöna.

4.2   Line R5 Praha - Ústí nad Labem - Cheb

Train paths of line R5 Praha - Ústí nad Labem - Cheb were constructed in 8 variants which are different:

- in reaching IPT-node Cheb (at minute 30 or at minute 00),
- in routing in Ústí nad Labem (via Ústí nad Labem hl.n. passenger station as in current state, or out of this station, which means directly to/from Ústí nad Labem západ station),
- in vehicles (using of current vehicles or using of vehicles with better dynamical properties).

Current vehicles are represented by an electric engine of class 151 in section Praha - Ústí nad Labem, an electric engine of class 362 in Ústí nad Labem - Cheb section and by 6 carriages of classic construction. In case of routing out of Ústí nad Labem hl.n. passenger station it is considered using of an electric engine of class 362 in whole Praha - Cheb section. As a train set with better dynamical properties five-part EMU of class 660 "InterPanter" was selected. Figure 1 schematically displays an overview of all the variants of constructed line R5 train paths.

Train Path CH30-HL. Train path CH30-HL of line R5 reaches IPT-node Cheb at time position X:30. It is routed via Ústí nad Labem hl.n. passenger station. In direction from Prague to Cheb the construction of the path follows these points (in the opposite direction it is constructed symmetrically):

- pass through Lovosice station at the latest minimum departure headway before departure of a train of line U4 in direction Ústí nad Labem,
- pass through Oldřichov u Duchcova station at the earliest minimum arrival headway after arrival of a train of line U24 in direction Louka u Litvínova,
- arrival at Klášterec nad Ohří station at the earliest minimum arrival headway after arrival of a train of line U2 from Most.

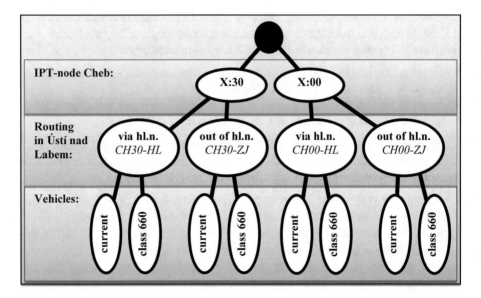

Fig. 1. Schematic overview of variants of constructed line R5 train paths

Train Path CH30-ZJ. For train path CH30-ZJ the same construction conditions are applied as for train path CH30-HL. The only difference is in routing in Ústí nad Labem, train path CH30-ZJ is not routed via Ústí nad Labem hl.n. passenger station. Train Path CH00-HL. Train path CH00-HL of line R5 reaches IPT-node Cheb at minute 00. It is routed via Ústí nad Labem hl.n. passenger station. In direction from Prague to Cheb the construction of the path follows these points (in the opposite direction it is constructed symmetrically):

- arrival at Most station at the earliest minimum arrival headway after arrival of a train of line U1 from direction Děčín,
- departure at Most station at the latest minimum departure headway before departure of a train of line U1, or U2 in direction Chomutov.

Train Path CH00-ZJ. For train path CH00-ZJ the same construction conditions are applied as for train path CH00-HL. The only difference is in routing in Ústí nad Labem, train path CH00-ZJ is not routed via Ústí nad Labem hl.n. passenger station.

4.3   Line R20 Praha - Ústí nad Labem - Děčín

For line R20 Praha - Ústí nad Labem - Děčín two train paths were constructed which vary according to used vehicles:

- current vehicles (an electric engine of class 162 and 7 carriages of classic construction),
- train set with better dynamical properties (three-part + five-part EMU of class 660).

With respect to determined restricting conditions (reaching of IPT-node Lovosice at X:00) only one train path for line R20 can be essentially constructed. Only few minutes time movements are permissible in order to keep transfer bindings at station Lovosice. Depending on used train sets (which have different runtimes) the difference in time position are only in Děčín - Lovosice (outside) and Lovosice (outside) - Praha sections.

5 Assessment of Constructed Train Paths

In this section the train paths constructed in Sect. 4 are evaluated, where mutual technological harmony between train paths of lines Ex3, R5 and R20 and harmony with regional lines as well are researched. The result of the assessment is elimination of such train paths, which are not convenient for further processing.

Train paths of line R20 are almost unequivocally given, because they are fixed by reaching IPT-node X:00 in Lovosice to reach transfer bindings to line U11 and other regional (bus) lines. For selection of inconvenient train paths of lines Ex3 and R5 following rules are used:

- 1 condition: if the train path involuntarily causes such time movement of line R20 train path which causes disruption of transfer bindings in IPT-node Lovosice, it is assessed as inconvenient and is not further used,
- 2 condition: if the train path involuntarily causes overtaking of line R20 train path or it is necessary to disproportionately increase runtimes of the path so that the line R20 train path would not be overtaken, it is assessed as inconvenient and is not further used,
- 3 condition: if the train path in combination with line R20 train path makes 30-min interval of line S4 in Kralupy nad Vltavou - Praha Masarykovo nádraží section impossible, or it is necessary to disproportionately increase runtimes or dwell times of the path to make it possible, it is assessed as inconvenient and is not further used,
- 4 condition: if the train path in combination with line R20 train path causes overtaking of line S4 train path in Kralupy nad Vltavou - Praha Masarykovo nádraží section or it is necessary to disproportionately increase runtimes or dwell times of the path so that the S4 train path would not be overtaken, it is assessed as inconvenient and is not further used,
- 5 condition: If the train path in combination with line R20 train path causes disruption of line S4/U4 integrity (thus, dwell time at least at one of the intermediate stations/stops is involuntarily lengthened - dwell times lower or equal to 10 min are tolerated in case when the train is overtaken) or it is necessary to disproportionately increase runtimes or dwell times of the path so that the integrity of line S4/U4 would not be disrupted, it is assessed as inconvenient and is not further used.

Comment to condition No. 5: when S4/U4 reaches IPT-node Lovosice at X:30 and the integrity of this line is required it is necessary to overtake this line by line R20 at Kralupy nad Vltavou station in order to keep transfer bindings between R20 and S4/U4 (identically with the current state).

5.1   Paths of Line Ex3

Train paths DR00-A and DR30 of line Ex3 were recommended for further use. Train path DR00-B was eliminated as an inconvenient one, because overtaking of line S4 by the train path DR00-B would be necessary at a station between Kralupy nad Vltavou and Praha-Holešovice-Stromovka (the 4. condition) or the integrity of line S4 would be disrupted by ca 15 min dwell time at Kralupy nad Vltavou station (the 5. condition), see Fig. 2.

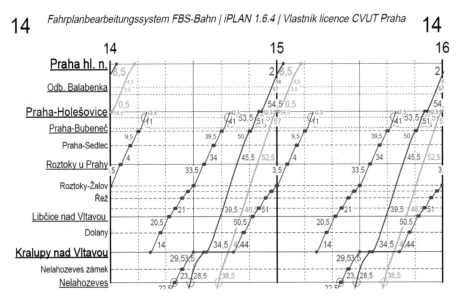

Fig. 2. Trains of lines R20, Ex3 (train path DR00-B) and S4 sequence in Nelahozeves - Kralupy nad Vltavou - Praha hl.n. section. Overtaking of the train of line S4 at Kralupy nad Vltavou station by the train of line R20 even by the train of line Ex3. Green color = line Ex3, brown color = line R20, blue color = line S4

5.2   Paths of Line R5

All train paths which reach IPT-node Cheb at time position X:30, namely CH30-HL and CH30-ZJ, are recommended for further use. Train paths which reach IPT-node Cheb at time position X:00, namely CH00-HL and CH00-ZJ, were eliminated as inconvenient ones - because of their construction conditions on railway lines 130 and 140 a movement of train paths of line R20 is caused which disrupts reaching of IPT-node Lovosice at X:00 (the 1st condition), see Figs. 3 and 4, or it is necessary to overtake trains of line R20 in an intermediate station between Ústí nad Labem and Prague (the 2nd condition).

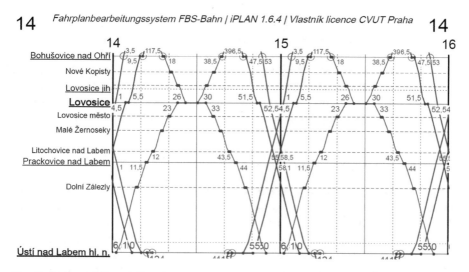

Fig. 3. Trains of lines R5 (train path CH00-HL) and R20 sequence in Ústí nad Labem - Lovosice - Bohušovice nad Ohří section. Inconvenient reaching of IPT-node Lovosice by line R20. Red color = line R5, brown color = line R20, blue color = line U4

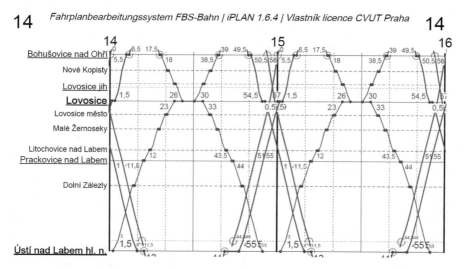

Fig. 4. Trains of lines R5 (train path CH00-ZJ) and R20 sequence in Ústí nad Labem - Lovosice - Bohušovice nad Ohří section. Inconvenient reaching of IPT-node Lovosice by line R20. Red color = line R5, brown color = line R20, blue color = line U4

6 Proposal of Operational Conceptions

Such train paths, which were selected as convenient for further use, were applied for proposal of two different operational conceptions. In both concepts the train paths of lines R5 and R20 are constructed for EMU of class 660. Situation in Prague node itself was not solved during creating of the new operational conceptions - critical places are especially junction Balabenka, section between junction Balabenka - Praha hl.n. and Praha hl.n. itself (technological intervals and track occupation at the station) [11].

It is important to take into consideration during creating and appraisal of the conceptions, that the evener layout of transport supply is used, the bigger profit for served relations exists. This means that variants where train paths of researched long-distance lines are less bundled between Praha and Ústí nad Labem are more attractive for passengers.

6.1   Operational Conception - Variant "Bundle"

In the first variant of proposed operational conception are train paths of line Ex3 and R5 in not tight bundle. There are used train path DR00-A of line Ex3 and train path CH30-HL of line R5. Bundle of these lines is not very convenient from transportation viewpoint, because there are offered two connections of express segment in Ústí nad Labem - Praha relation in relatively short time space, but then almost 60-min space follows to next connection of express segment (which is interposed only with slower connection of the second transportation segment - by line R20). For higher trans-portation attractivity, there is an effort to make the bundle of lines Ex3 and R5 as wide as possible, therefore overtaking of line S4 by these lines is realized at two stations (at Lovosice by line Ex3, at Bohušovice nad Ohří by line R5).

A great advantage of this operational conception is a possibility of early intro-duction, because current IPT-nodes Cheb (X:30) and Dresden (X:00) are kept. The netgraph of this operational conception is displayed in Fig. 5.

6.2   Operational Conception - Variant "Interposition"

In the second variant of proposed operational conception, there are the train paths of lines Ex3 and R5 interposed between Praha hl.n. and Ústí nad Labem hl.n. into common 30-min interval. There are used train paths DR30 of line Ex3 (IPT-node Dresden at X:30) and CH30-HL of line R5 (IPT-node Cheb at X:30).

The solution with interposition of express-train paths is better from the trans-portation viewpoint, because there is offered more attractive supply in relation Ústí nad Labem - Prague. With respect to travel time Dresden Hbf - Ústí nad Labem hl.n. of line Ex3 and travel time Cheb - Ústí nad Labem hl.n. of line R5, there is also possible to create a secondary IPT-node in Ústí nad Labem hl.n. at time position X:15/X:45 with a "corner" transfer binding for journeys between Dresden - Ústí nad Labem - Cheb. Disadvantages of this operational conception are:

- line R20 reaches the IPT-node Lovosice at X:00 not sharply, so the transfer bindings to line U11 are very tight and essentially without time reserves (but minimal changing times are observed),
- there is no regular connections supply between Ústí nad Labem hl.n. - Děčín hl.n., where connections of lines U1, R20, Ex3 (alternatively R15 too) depart in a 30-min time space and during the following 30-min there is no connection. Short time spaces between trains of lines R20 and Ex3 could be solved by shortening line R20 only to relation Praha hl.n. - Ústí nad Labem hl.n.

The netgraph of this operational conception is displayed in Fig. 6.

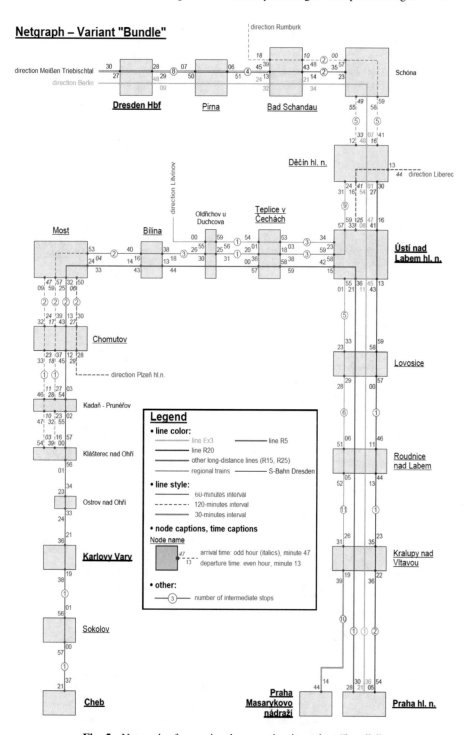

Fig. 5. Netgraph of operational conception in variant "bundle"

Netgraph – Variant "Interposition"

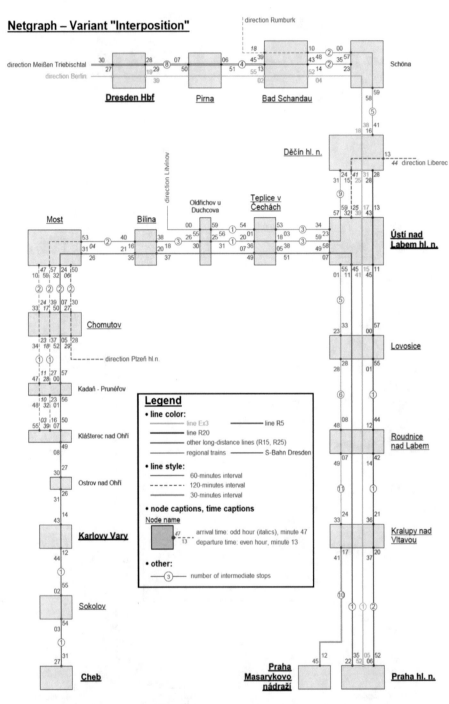

Fig. 6. Netgraph of operational conception in variant "interposition"

7 Conclusion

This paper described, how possibilities of new conception of long-distance railway transport in area Praha - Dresden/Cheb were tested and created. At first there was made an analysis of current operational conception of long-distance lines Ex3 Praha - Ústí nad Labem - Děčín, R5 Praha - Ústí nad Labem - Cheb and R20 Praha - Ústí nad Labem - Děčín. The general analysis was followed by an analysis of transportation relations in the researched area - for both domestic and cross-border transportation relations. For important long-distance transport relations, the competitiveness of railway transport was tested (as a comparison of travel times to alternative transport modes). The analysis clearly showed, that 60-min interval (at least during peak hours) on both lines Ex3 (at least in Praha - Děčín section, ideally even to Berlin) and R5 (at least in Praha - Chomutov section) would respond to potential passenger demand and is reasonable from the transportation viewpoint (even on current railway infrastructure).

For proposing of changes of the operational conception, there was necessary to formulate the restricting technological conditions for construction of the researched lines train paths. Besides technological research the network bindings and reaching of IPT-nodes had to be considered, not only in case of researched long-distance lines (line Ex3 IPT-node Dresden, line R5 IPT-node Cheb, line R20 IPT-node Lovosice) but in case of related lines as well. Necessity of keeping these network bindings represented a significant boundary condition for reaching a suitable solution.

Based on the determined restricting conditions possible variants of train paths of the researched long-distance lines were constructed. Overall 3 train paths of line Ex3, 8 train paths of line R5 and 2 train paths of line R20 were constructed. Then these train paths were evaluated. During the evaluation there was tested the coordination between trains paths of lines Ex3, R5, R20 and S4/U4. The result was, that 1 train path of line Ex3 and 4 train paths of line R5 were evaluated as inconvenient and were eliminated.

The other train paths were then used for proposal of two possible (from transportation even from operational viewpoints) operational conceptions, both respected different time positions of IPT-nodes and different mutual interposition of long-distance lines.

The first operational conception was proposed for keeping the current time positions of IPT-nodes Dresden (X:00) and Cheb (X:30). Train paths of lines Ex3 and R5 get into a (not tight) bundle between Ústí nad Labem – Praha. The second operational conception was proposed to reach IPT-node Dresden by line Ex3 at X:30. In this variant the train paths of lines Ex3 and R5 are interposed into their common 30-min interval.

The main aims of the conceptual proposal were fully achieved. Both found final variants of a new transport concept are technologically realizable and permit extension of operation corresponding to transportation flows in the researched area. Found solution provides such selection of the preferred variant, which is convenient from required network bindings viewpoint.

References

1. Ministerstvo Dopravy ČR: Plán dopravní obslužnosti území vlaky celostátní dopravy: zásady objednávky dálkové dopravy pro období 2017–2020. https://www.mdcr.cz/getattachment/Dokumenty/Verejna-doprava/Financni-ucast-statu/Plan-dopravni-obsluhy-uzemi-vlaky-celostatni-dopra/Plan-dopravni-obsluhy-uzemi-2017-2021.pdf.aspx
2. SŽDC: Rail Operation. https://www.szdc.cz/en/provozovani-drahy.html
3. Czech Statistical Office. Population and Housing Census 2011. https://www.czso.cz/csu/czso/population-and-housing-census
4. CHAPS: Search for connections. https://jizdnirady.idnes.cz/?lng=E
5. Seznam.cz. Route Planner. Mapy.cz. https://en.mapy.cz/zakladni?planovani-trasy
6. European Commission: Joint Research Centre, Institute for Prospective Technological Studies. Transtools. http://energy.jrc.ec.europa.eu/transtools/index.html
7. Janoš, V., Kříž, M.: Using of the EVA model in the Czech Republic. In: Road and Rail Infrastructure V. Proceedings of the Conference CETRA 2018, pp. 1369–1376. University of Zagreb, Zagreb (2018)
8. Janoš, V., Kříž, M.: Pragmatic approach in regional rail transport planning. Sci. J. Sil. Univ. Technol. Ser. Transp. **100**, 35–43 (2018)
9. Drábek, M., Pospíšil, J.: Fluctuations in passenger railway service period. In: Young Transportation Engineers Conference 2018, pp. 1–8. Czech Technical University in Prague, Faculty of Transportation Sciences, Prague (2018)
10. Bundesministerium für Verkehr und Digitale Infrastruktur. Infrastruktur für Einen Deutschland-Takt im Schienenverkehr. BMVI (2019). https://www.bmvi.de/goto?id=316588
11. Drábek, M., Janoš, V., Michl, Z.: Quantitative determination of bottlenecks in railway networks with periodic service. In: Proceedings of 20th International Conference Transport Means 2016, pp. 594–598. Kauno Technologijos Universitetas, Kaunas (2016)

The Concept of Rules and Recommendations for Riding Shared and Private E-Scooters in the Road Network in the Light of Global Problems

Katarzyna Turoń[(⊠)] and Piotr Czech

Faculty of Transport, Silesian University of Technology, Katowice, Poland
{katarzyna.turon, piotr.czech}@polsl.pl

Abstract. The work was dedicated to the topic of riding e-scooters in urban transport systems. These kind of new mobility solution is becoming more and more popular in the case of shared mobility operators and also the new social trend. It has been noticed by the authors that the appearance of a large number of e-scooters in urban traffic conditions, has caused many problems related to safety and legal issues. That problems occur in issues related to moving and park e-scooters on the streets, intersections as well as pavements. Because of that fact, the authors tried to indicate conditions for flowing movement of e-scooters in the city. The aim of the work was to present the concept of rules and recommendations for riding individual and shared e-scooters in the road net-work. In addition, the text also gives global examples of how e-scooters operate in urban transport systems.

Keywords: E-scooter-sharing · Scooter-sharing · E-scooters ·
Micro-mobility · E-scooter-sharing · Legal regulations · Safety of e-scooter
sharing · E-scooter-sharing and infrastructure · Shared mobility problems ·
Personal transporter

1 Introduction

Shared mobility solutions are one of the possibilities to implement sustainable transport development policy in practice in urban logistics. Initially, the public was skeptical about the sharing of vehicles [1], explaining the lack of a vehicle as a loss of prestige or disruption of their current lifestyle [2]. But over time, people began to learn about shared vehicles [3], implementing in their habits the so-called 'new culture of mobility' [4] and in some cases even decided to give up buying a second car in the family or to stop using it for shared mobility services [5]. This kind of behavior involved the trend of the economy of collaboration [6]. It provided shared mobility to becoming more and more popular around the world. Firstly, in urban logistics bike-sharing systems were occurred. And now, the most popular part of shared mobility is e-scooter-sharing (e-kick-scooter-sharing) [7]. The emergence of many electric e-scooter rentals has started the trend of buying e-scooters for their own use by individual users. As a result, several thousand e-scooters from rentals with indefinite number of private e-scooters have

E. Macioszek and G. Sierpiński (Eds.): Modern Traffic Engineering in the System Approach
to the Development of Traffic Networks, AISC 1083, pp. 275–284, 2020.
https://doi.org/10.1007/978-3-030-34069-8_21

appeared in the road traffic. Due to the development of e-scooter-riding trend, also many controversies have been appeared. Mainly these controversies are related to the relevant legislation, driving rules, e-scooters, aspects related to safety as well as chaos that e-scooters cause in cities [8]. Due to the above issues, the authors decided to dedicate this article to e-scooter sharing systems.

The aim of the work was to present the concept of rules and recommendations for riding e-scooters in the road network in the light of global examples. In the work was shown the idea of shared mobility in the case of e-scooters with its advantages and disadvantages. In the next part of the article examples of e-scooter-sharing systems and regulations were presented. In the last part of work was presented the concept of recommendations for safe and fluent riding e-scooters from e-scooters-sharing services.

The article is a proposal for the scientific niche connected with the lack of the knowledge related to the e-scooter-sharing conception. It can be used as a support for global authorities to implement new conditions dedicated to riding e-scooters.

2   Idea of E-Scooter-Sharing and E-Scooter Riding

The scooter sharing/scooter-sharing (properly e-scooter sharing) system is a service in which e-scooters are made available for use in short-term rentals. The functioning of scooter sharing is similar to bike-sharing or car-sharing systems [5]. Some scooter companies offer more than one type of vehicle through their services. Scooter models differ mainly in type of construction, kilometer range and possible load capacity of the vehicle. Examples of e-scooters are presented in the Fig. 1.

Fig. 1. Example of e-scooters fleet in on the sidewalk in Paris, France

The idea of e-scooter-sharing is to using them as "dockless" mode of transportation. This means that the majority of vehicles do not have permanent rack/dock parking spaces. That kind of solution is perfect as in the case of bike-sharing of the dockless type in terms of accessibility for the user and the first and last transport mile [9].

Electric scooters have many benefits. The main ones include:

- providing an additional transport solution that gives users the opportunity to use during the first and last miles,
- additional transport solution ensuring touristic attractiveness of cities where electric scooters-sharing systems are located,
- positive impact on the environment thanks to electro-mobility and reduction of negative noise,
- education for electro-mobility thanks to the high availability of vehicles throughout the whole society,
- relatively low cost of buying the scooter,
- low costs of vehicle maintenance and upkeep.

Of course, as with any mode of transport, electric scooters also have disadvantages. The main ones include:

- low speed of movement in relation to i.e. car or bus,
- low battery range,
- the need to charge the vehicle,
- it is not possible to transport more than one person,
- problematic issue of luggage transport.

The first scooter-sharing system launched in San Francisco since the end of 2011 [10], but in 2017–2018 they showed their greatest development. At the end of 2018, a total of 85,000 e-scooters were available in nearly 100 cities in the United States, which were rented 38.5 million times last year [11]. E-scooters belonged to just 4 of U.S.: California, Texas, Florida and Massachusetts [12].

In the case of Europe, the largest increase in interest in scooter-sharing services occurred in 2018, when he systems began to operate in the biggest European capitals. Statistics related to selected system implementations in individual countries in subsequent years are presented in Table 1.

Table 1. Selected e-scooter-sharing in the world - statement (Source: author's own collaboration based on: [12–16]).

Country	City	Operator	Year	Funding
U.S.	San Francisco	Scoot Networks	2011	$ 46 million
U.S.	San Francisco	Lime	2017	$ 765 million
U.S.	Santa Monica	Bird	2017	$ 415 million
Mexico	Mexico City	Grow mobility	2018	$ 150 million
Germany	Berlin	Flash	2018	$ 66 million
Sweden	Stockholm	Voi Technology	2018	$ 47 million
U.S.	San Francisco	Skip	2018	$ 31 million
Netherlands	Amsterdam	Dott	2018	$ 23 million
U.S.	San Antonio	Blue Duck	2018	$ 23 million
Germany	Berlin	Wind Mobility	2018	$ 22 million

The electric scooter-sharing systems in other countries are also rapidly developing. There are currently 12 e-scooter-sharing system operators in Paris, in which over 20,000 vehicles are available [17]. In the case of Poland, there are currently 10 scooter rental operators. The first of the systems began its operations in 2018. Currently, from shared e-scooters - in total almost 8 thousand vehicles - residents of 10 Polish cities can benefit [11].

The occurrence of systems of shared electric mobility scooters has caused a lot of interest in having individual users scooters on the property. This type of behavior is mainly due to economic reasons due to the relatively high costs of renting vehicles in sharing services. As a result, in addition to scooters from the sharing systems, there are also many individual scooters on the streets. These kind of individual means of transport is called "personal transporter". The personal transporter is a class of compact, modern and motorized vehicle for transporting an individual at speeds that do not normally exceed 25 km/h They include electric skateboards, kick scooters, self-balancing unicycles and Segway's, as well as gasoline-fueled motorized scooters or skateboards [18, 19]. The presence of such a large number of scooters in street traffic caused communication chaos. It would be worth carrying out various types of traffic measurements [20–22] to find out the exact state of safety in the operation of scooters. Therefore, it is important to focus on the appropriate rules that should be taken into account when riding scooters.

3   Examples of Problems Related to Moving Scooters in Road Traffic

The presence of a large number of electric scooters in city traffic has led to numerous safety and chaos problems. First of all, scooters with the option of leaving them anywhere became a communication barrier for other traffic users [23]. It is worth mentioning here pedestrians or people with disabilities who are exposed to problems communicating around the city in connection with parked on the sidewalks and nearby or at scooters intersections. An example of leaving scooters anywhere in the city is shown in Fig. 2.

Another example is non-compliance with safety rules when moving with scooters. This means, for example, riding a scooter in two, driving on pavements and the like. Example is presented in Fig. 3.

There is also a problem of abandoned equipment in the city. Scooters are devastated, thrown into rivers or their equipment is stolen.

In addition, it is worth mentioning that currently hospitals are facing a very big problem related to people who are victims of a collision or accidents related to riding a scooter. Scooter users hover over pedestrians or prevent them from moving on the sidewalks. Fatal accidents have also been reported.

There are also legislative problems related to the difficulty of matching the scooter to current regulations. It is noteworthy that the scooter boom came about when in many countries no law was developed. In addition, it was not specified whether scooters can be used on pavements, roads or bicycle paths and what the electric scooter exactly is. It is also worth mentioning the problems with liability insurance in the case of scooters and the lack of appropriate regulations related to the need to ride in a helmet or without it.

Fig. 2. The example of leaving scooters dockless - anywhere in the city in Brussels, Belgium

Fig. 3. Example of improper driving of the electric scooter by two people in Brussels, Belgium

What is more there is also a problem related to the ecological issue of users dropping scooters. In many cases, scooters are not picked up by operators because it is unprofitable for them. As a result, they are another waste poisoning the natural environment.

In addition, there is the problem of liability for a collision or accident. He is not entirely sure who should be responsible for what type of offense he should be and what the amount of punishment should be. There are also no signs dedicated to scooters as it is in the case of bicycles.

4   The Concept and Recommendation Related to Riding E-Scooters

Due to the many problems associated with moving scooters in urban conditions, a set of rules and recommendations have been developed to improve traffic and safety. It is worth to:

- adjust the law that will take into account the exact definition of the e-scooter and will determine its allowable speed, equipment and places where it can be navigated. It is also worth defining whether the scooter should undergo technical tests, what the process of putting it into service should look like and what are the requirements for its inspections,
- it is worth introducing legal provisions regarding leaving the scooter freely in the public space. Specially dedicated docking stations may be the solution. Figure 4 shows an example of a docking station, which is also a bench and does not disturb architectural order. It is also worth developing a concept of parking spaces dedicated to scooters,

Fig. 4. The example of a functional docking station for electric scooters in Valencia, Spain (Source: shared by Mr. Santiago Jose Hernandez Bethencourt CEO in the company Scooterdocks)

- user security policies should be defined. It is worth determining whether users who ride scooters should wear helmets or additional protectors,
- it is worth informing clearly about issues related to the type of liability insurance and accident insurance. Do operators provide them or should responsible users purchase policies on their own,
- specify age regulations for users who want to ride electric scooters. In addition, specify whether you need a bicycle card or driving license for this entitlement,
- rules should be laid down for the use of electric scooters by intoxicated drivers or those under the influence of drugs,
- appropriate tariffs for fines for irregularities related to scooters should be introduced,
- cities should consider the principles of cooperation with operators. Currently, the freedom of the scooter market is leading to a very large communication chaos. Appropriate city-operator cooperation and limiting the market to a few operators, not a dozen or so, will give you the opportunity to control the excessive share of scooters in the modal division,
- cities should know the scooter rental details - the data. So to properly control and manage the number of scooters available to users in given places,
- it is also worth constantly monitoring the location of scooters and their parking places. They avoided the devastation of vehicles,
- take all kinds of educational activities in the field of electro-mobility and new mobility, which will explain to society how to safely use modern means of transport. This can be achieved by creating educational portals or creating international projects devoted to the topic of new mobility [24, 25].

5   E-Scooter Policies - World Examples

The problem of scooters has led to the development of special types of policies and regulations dedicated to people using these devices. It is worth mentioning that riding electric scooters in the United Kingdom is prohibited to ride on the streets and pavements [26]. The only place where riding is possible is private land, with the consent of the owner [26]. Riders currently face a £300 fixed-penalty notice and six points on their driving license for using e-scooters [26]. In the UK, electric scooters are treated as Personal Light Electric Vehicles (PLEVs), which means they are motor vehicles [26]. That means they are subject to all the requirements a motor vehicle is subject to - MOT, tax, licensing and construction requirements - such as having visible rear red lights, number plates and signaling ability [26]. Due to the fact that scooters do not have this type of authorization, they are not allowed to be used in the territory of the UK [26].

In the case of France there is no permission to riding on pavements, except for pushing the scooter with the engine off [27]. The maximum speed is 25 km/h [25]. The scooter can only be used on bicycle paths or roads where the maximum speed is 50 km/h. E-scooters can be park only in special designated areas [28]. It is not allowed to transport more than one person on board the device [27].

High penalties related to non-compliance with the law have been introduced [27]:

- siding on pavements - penalty of €135,
- parking in a prohibited place - penalty €35.

In the case of Austria electric scooters have the permission to developing speeds of up to 25 km/h [27]. It is allowed to riding in bicycle paths [28]. E-scooters are subject to the same speed limits and traffic laws, including driving on pavements and the obligation to equip the vehicle with brakes, bell and lighting [28].

In the case of Germany all of e-scooters need to have traffic authorization [27]. It is possible to ride only on bicycle routes, including counter-lanes [29]. The maximum speed of the e-scooter is 20 km/h [29]. The user need to have compulsory liability insurance, compulsory equipment: handbrake and lighting. There is no obligation to drive in a helmet [29].

In the case of United States there are separate regulations depending on the state. In the California the user must have a valid driver's license or learners permit [30]. It is ordered to ride in a helmet. E-scooters can ride with maximum speed - 15 mph. It is not allowed to ride on the pavement. Users can riding on bicycle paths [30].

In Texas allow e-scooters on sidewalks and on streets with a speed limit under 35 mph [30]. And electric scooters use the same traffic laws that apply to bicycles such as obeying speed limits and using your arms to signal turns [30]. In the New York doesn't allow "motorized scooters" on any street, highway, parking lot or sidewalk [30].

6 Summary

In summary, in the era of increasing attention to education for electro-mobility [31] shaping appropriate transport behavior [32, 33] and the creation of various types of portals aimed at raising awareness of its positive sides for society, shared mobility solutions will become increasingly popular. electric scooters are the current trend in new mobility services. They certainly make it easier for users to move around the city and constitute an additional transport alternative. However, with all its advantages, there are also security risks. Therefore, the authors of this article pointed out what should focus the laws governing the rules of riding scooters in urban traffic. The authors tried to take into account both legal, safety and infrastructure issues The developed concept of law as well as recommendations can provide support when trying to regulate the market of electric scooters, both from rental and private. In further research, the authors plan to analyze compliance with the law related to electric scooters in various countries.

References

1. Currie, G.: Lies, damned lies, AVs, shared mobility, and urban transit futures. J. Public Transp. **21**(1), 19–30 (2018)
2. Bryman, A., Gillingwater, D., Warrington, A.: Community-based transport coordination strategies: the UN Experiences. In: Gillingwater, D., Sutton, J. (eds.) Transportation Studies. Community Transport Policy, Planning, Practice, vol. 15, pp. 117–140. Routledge, Taylor & Francis Group, London and New York (2014)
3. Bryman, A., Gillingwater, D., McGuiness, I.: Decision-making process in community transport organisations. In: Gillingwater, D., Sutton, J. (eds.) Transportation Studies. Community Transport Policy, Planning, Practice, vol. 15, pp. 141–160. Routledge, Taylor & Francis Group, London and New York (2014)
4. Sierpiński, G., Staniek, M., Celiński, I.: New methods for pro-ecological travel behavior learning. In: Gómez Chova, L., López Martínez, A., Candel Torres, I. (eds.) ICERI 2015 Proceedings of 8th International Conference of Education, Research and Innovation, pp. 6926–6933. IATED, Sevilla (2015)
5. Turoń, K., Czech, P.: Polish systems of car-sharing - the overview of business to customer service market. In: Macioszek, E., Sierpiński, G. (eds.) Directions of Development of Transport Network and Traffic Engineering. LNNS, vol. 51, pp. 17–26. Springer, Cham (2019)
6. Chris, J.M.: The sharing economy: a pathway to sustainability or a nightmarish from of neoliberal capitalism? Ecol. Econ. **121**, 149–159 (2016)
7. McKenzie, G.: Spatiotemporal comparative analysis of e-scooter-share and bike-share usage patterns in Washington, D.C. J. Transp. Geogr. **78**, 19–28 (2019)
8. Allem, J.P., Majmundar, A.: Are electric e-scooters promoted on social media with safety in mind? A case study on bird's instagram. J. Prev. Med. Public Health **13**, 62–63 (2019)
9. Chen, F., Turoń, K., Kłos, M.J., Czech, P., Pamuła, W., Sierpiński, G.: Fifth generation of bike-sharing systems - examples of Poland and China. Sci. J. Sil. Univ. Technol. Ser. Transp. **99**, 5–13 (2018)
10. Zipcar For Scooters. https://www.zipcar.com/san-francisco?&gclid=Cj0KCQjw-b7qBRDP ARIsADVbUbV65zTrCsC3ttaGqd9YHwmSczYi_-_pTI0Y1sbEnStXnKkUST5-WfAaAos NEALw_wcB
11. Mashable Portal. https://mashable.com/article/lime-escooters-growth-us-global/?europe= true. E-scooters aren't going anywhere — in fact, their numbers are still growing
12. Mobility Foreshights, Electric Scooter Sharing Market in US and Europe 2019–2024. https:// mobilityforesights.com/product/scooter-sharing-market-report/
13. Crunchbase Portal. https://www.crunchbase.com/lists/electric-scooter-and-ride-sharing/ 72a32e46-452e-483d-9d03-99362a2030 f0/organization.companies
14. Pitchbook Portal. https://pitchbook.com/newsletter/sounding-off-on-the-future-of-electric-scooters
15. TechCrunch. https://techcrunch.com/2018/12/23/the-electric-scooter-wars-of-2018/
16. BCG Analysis. https://www.bcg.com/publications/2019/promise-pitfalls-e-scooter-sharing. aspx
17. France 24 Portal. https://www.france24.com/en/20190607-france-paris-cracks-down-electric-scooters-anarchy-mayor-hidalgo-pavements-parking
18. Information Sheet - Guidance on Powered Transporters. Department for Transport. https:// assets.publishing.service.gov.uk/government/uploads/system/uploads/attachment_data/file/ 467289/motorcycle-retailers-rules-january-2016.pdf

19. Litman, T.: Managing diverse modes and activities on nonmotorized facilities: guidance for practitioners. ITE J. **76**(6), 20–27 (2006)
20. Macioszek, E., Lach, D.: Comparative analysis of the results of general traffic measurements for the Silesian Voivodeship and Poland. Sci. J. Sil. Univ. Technol. Ser. Transp. **100**, 105–113 (2018)
21. Macioszek, E., Lach, D.: Analysis of traffic conditions at the Brzezinska and Nowochrzanowska intersection in Myslowice (Silesian Province, Poland). Sci. J. Sil. Univ. Technol. Ser. Transp. **98**, 81–88 (2018)
22. Macioszek, E.: Changes in values of traffic volume - case study based on general traffic measurements in Opolskie Voivodeship (Poland). In: Macioszek, E., Sierpiński, G. (eds.) Directions of Development of Transport Networks and Traffic Engineering. LNNS, vol. 51, pp. 66–76. Springer, Cham (2019)
23. Macioszek, E.: The comparison of models for critical headways estimation at roundabouts. In: Macioszek, E., Sierpiński, G. (eds.) Contemporary Challenges of Transport Systems and Traffic Engineering. LNNS, vol. 2, pp. 205–219. Springer, Cham (2017)
24. Sierpiński, G., Staniek, M.: Education by access to visual information - methodology of moulding behaviour based on international research project experiences. In: Gómez Chova, L., López Martínez, A., Candel Torres, I. (eds). ICERI 2016 Proceedings of 9th International Conference of Education, Research and Innovation, pp. 6724–6729. IATED Academy, Sevilla (2016)
25. Sierpiński, G., Staniek, M., Celiński, I.: Travel behavior profiling using a trip planner. Transp. Res. Procedia **14C**, 1743–1752 (2016)
26. BBC News Portal. https://www.bbc.com/news/uk-48106617
27. France News Portal. http://en.rfi.fr/france/20190504-new-french-law-ban-electric-scooters-pavements
28. Government of Austria Portal. https://www.loc.gov/law/foreign-news/article/austria-ban-on-riding-e-scooters-on-sidewalks-enacted/
29. The Local Deutschland Portal. https://www.thelocal.de/20190517/e-scooters-get-the-green-light-on-germanys-roads
30. Swagtron Portal. https://swagtron.com/news/are-electric-scooters-legal/
31. Sierpiński, G., Staniek, M.: Platform to support the implementation of electromobility in smart cities Based on ICT applications - concept of electric travelling project. Sci. J. Sil. Univ. Technol. Ser. Transp. **100**, 181–189 (2018)
32. Staniek, M., Sierpiński, G., Celiński, I.: Shaping environmental friendly behaviour in transport of goods - new tool and education. In: Gómez Chova, L., López Martínez, A., Candel Torres, I. (eds.) ICERI 2015 Proceedings of 8th International Conference of Education, Research and Innovation, pp. 118–123. IATED Academy, Seville (2015)
33. Sierpiński, G., Celiński, I., Staniek, M.: The model of modal split organisation in wide urban areas using contemporary telematic systems. In: Proceedings of the 3rd International Conference on Transportation Information and Safety, pp. 277–283. IEEE Press, Wuhan (2015)

Video-Based Distance Estimation for a Stream of Vehicles

Zbigniew Czapla$^{(\boxtimes)}$

Faculty of Transport, Silesian University of Technology, Katowice, Poland
zbigniew.czapla@polsl.pl

Abstract. A novel measuring method of a distance between vehicles in a vehicle stream is presented in the paper. The presented method utilizes vision data for vehicle detection. Input images are converted into binary images, Conversion is performed with the use of analysis of small gradient magnitudes in the input images. Two detection fields, initial and final, are defined. Vehicle detection is carried out by analysis of the state of detection fields. The changes of the state of both detection fields, caused by passing vehicles, are indicate by appropriate image ordinal numbers from the sequence of input images. Application of two detection fields allows determination of a time period between successive vehicles and vehicle speed. The distance between two consecutive vehicles is calculated as a relation of ordinal image numbers. Processing quantities and experimental results are provided.

Keywords: Vehicle detection · Image gradients · Distance between vehicles · Vehicle speed

1 Introduction

Present-day road traffic systems employ vision data for traffic measurements and surveillance. Vision data are processed and analysed for vehicle detection and counting, division of vehicles into categories, and determination of vehicle trajectories. Vision date also allow estimation of other traffic parameters as speed of vehicles and a distance between successive vehicles.

Determination of traffic parameters on the basis of vision date requires vehicle detection. Description of trajectories of vehicles provides data for further calculations of individual traffic parameters. In [1] a system for determination of vehicles trajectories is presented. Vehicle detection and tracking is performed on the basis of feature analysis in regions where brightness changes in a few directions. Vehicles trajectories are determined by analysis of corner features. Traffic parameters are obtained from vehicles trajectories. In [2] traffic parameters are determined by application of static and dynamic analyses. Static analysis is employed to single images, in dynamic analysis preceding images also are considered. Vehicle tracking system for traffic monitoring at intersections is discussed in [3]. Processed images are divided into square blocks and analyzed with application of stochastic methods. In [4] segmentation is applied for vehicle tracing. Processing employs differences between a current image and a background image. The applied algorithm includes thresholding, bianarization and

E. Macioszek and G. Sierpiński (Eds.): Modern Traffic Engineering in the System Approach
to the Development of Traffic Networks, AISC 1083, pp. 285–295, 2020.
https://doi.org/10.1007/978-3-030-34069-8_22

background updating. In [5] vehicles are detected with the use of difference images. The presented method employs detection of lane dividing lines, elimination of shadows and Kalman filtering. Utilization of low-level modules and high-level modules for vehicles detection and tracking presents [6]. Low-level modules are destined for extracting image features, and high-level modules are employed for tracking vehicles. In [7] vehicles detection is carried out with the use of time-spatial images which are created on the basis of virtual lines. The virtual lines are defined in input frames and correspond to detected vehicles. In [8], an efficient method of background determination is presented. This method uses a mixture of Gaussian functions for pixel modelling. In [9] vehicles are considered as objects moving in one dimension. The presented method employs to speed estimation difference images and vehicle tracking based on calibration parameters of an applied camera. In [10], vehicle detection is carried out by frame differencing. Time relations and a speed of vehicles are determined on the basis of frame rate of a processed video sequence.

The presented method of distance estimation employs conversion of input images into binary images. Conversion into binary images is based on analysis of small gradient magnitudes in input images. Binary images are applied to detection of the vehicle passage through a detection field [11]. Estimation of the distance between vehicles involves determination of time relations and vehicle speed. Detection of vehicle speed can be performed by analysis of a sequence of binary images with defined two detection fields [12]. Assuming a known distance between detection fields, the distance between successive vehicles can be expressed as a function of frame numbers in an input video stream.

2   Input Data

Input data are obtained at a measuring station equipped with a video camera. The stationary video camera is located over a road and is directed towards approaching vehicles. The vehicles driving along the road and passing the measuring station are recorded by the video camera in a video stream. A passage of vehicles through the measuring station is registered in successive frames of the video stream. The diagram of the measuring station has been made. Figure 1 shows the diagram and demonstrates a vehicle passage through the measuring station. Three vehicle positions at the measuring station are marked in the diagram.

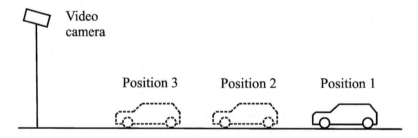

Fig. 1. Diagram of the measuring station

An input video stream obtained from the video camera consists of frames containing road scenes recorded at the measuring station. The consecutive frames of the input video stream constitute a sequence of input images. The sequence of input images consists of grey-scale images in the bitmap format. The test sequence of input images has been made at the test measuring station. Figure 2 shows the samples of input images presenting the same car passing the measuring station and recorded in different frames of video stream.

Fig. 2. Input images

Each input image consists of pixels located in N columns and M rows. In all images, the location of individual pixels is described by the column coordinate n, in the interval $[0, N - 1]$, and the row coordinate m, in the interval $[0, M - 1]$. A number of input images recorded during 1 s depends on a frame rate of the applied video camera. The frame rate is denoted by r and is expressed in frames per second. Each input image is labeled by the ordinal number denoted by i which indicates the position in the sequence of input images and corresponds to the frame number in the input video stream.

3 Binary Images

Input images are converted into binary images. Each image of the sequence of input images is processed separately. All images are represented by matrices. Individual input images are represented by input matrices $\mathbf{A_{(i)}} = [a_{(i)m,n}]$, and correlated with them binary images are represented by binary matrices $\mathbf{B_{(i)}} = [b_{(i)m,n}]$. All elements of the input matrix are equal to the values of appropriate pixels in the input image corresponding to the input matrix.

As a result of processing a single input image, all elements of the binary matrix are set to the binary value equal to 0 or 1. All non-border elements of input matrix are processed one by one in the growing in rows and columns order of matrix indices. In the input matrix, for each processed current element $a_{(i)m,n}$, small gradients in the neighbourhood are calculated. The small gradients are considered in rows, columns and two diagonal directions. In the binary matrix, the appropriate elements in the neighbourhood of the current element $b_{(i)m,n}$ are set to binary value equal to 1 on the basis of magnitudes of small gradients calculated for the input matrix and a preset threshold value T_a as follows:

$$b_{(i)m,n-1} = f\left(a_{(i)m,n}, \; a_{(i)m,n-1}, \; T_a\right),$$
$$b_{(i)m-1,n} = f\left(a_{(i)m,n}, \; a_{(i)m-1,n}, \; T_a\right),$$
$$b_{(i)m-1,n-1} = f\left(a_{(i)m,n}, \; a_{(i)m-,n-1}, \; T_a\right),$$
$$b_{(i)m+1,n-1} = f\left(a_{(i)m,n}, \; a_{(i)m+,n-1}, \; T_a\right). \tag{1}$$

The value of the current element $b_{(i)m,n}$ of the binary matrix, corresponding to the processed current element $a_{(i)m,n}$ of the input matrix, is established as the logical sum of the appropriate binary neighbourhood values and given by the equation.

$$b_{(i)m,n} = b_{(i)m,n-1} \vee b_{(i)m-1,n} \vee b_{(i)m-1,n-1} \vee b_{(i)m+1,n-1}. \tag{2}$$

The border elements of binary matrix (in rows at $m = 0$ or $m = M - 1$, and in columns at $n = 0$ or $n = N - 1$) are preset to binary 1. Test input images have been converted into binary images. The samples of binary images, corresponding to the samples of input images, are shown in Fig. 3. Black points in the binary images indicate the binary matrix elements equal to 1.

Fig. 3. Binary images

The elements of the binary matrices of a binary value at 1 correspond to edges of objects in input images. Hence, these elements have been called the edge elements. Moving objects in the input images can be detected by analysis of a number of edge elements in the binary images within areas defined as the detection fields.

4   Detection Fields

There is a pair of detection fields defined for each considered road lane. The first detection field has been named the initial detection field, similarly, the second detection field has been is named the final detection field. Both detection fields are situated along the considered road at a constant distance. For approaching vehicles, the initial detection field precedes the final detection field. The width of the detection fields corresponds to the width of a considered road lane. The detection fields have been marked in the diagram of the measuring station. Figure 4 shows a location of the detection fields and positions of a vehicle moving through the measuring station.

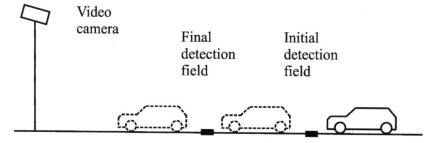

Fig. 4. Location of the detection fields in the measuring station

Each detection field is described by 4 parameters: upper row m_U, bottom row m_B, left column n_L and right column n_R. Hence, the initial detection fields is described by the sets of parameters $\{m_{iniU}, m_{iniB}, n_{iniL}, n_{iniR}\}$, and the final detection field by the set of parameters $\{m_{finU}, m_{finB}, n_{finL}, n_{finR}\}$.

In Figs. 5 and 6 the examples of input and binary images presenting a vehicle passage through the detection fields are shown respectively. In these images, the detection fields are indicated by black rectangles, Black points in the binary images indicate the edge elements.

Fig. 5. Detection fields in the input images

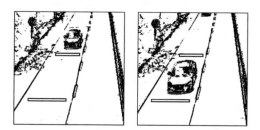

Fig. 6. Detection fields in the binary images

In binary images, the averaging numbers of the edge elements are determined separately inside the initial and final detection fields. The averaging number of the edge elements is computed for the current image denoted by i and the preset number of the previous images denoted by P as the average value of the number of edge elements inside the detection field. For the current image, the averaging number of the edge elements in the initial detection field is given by:

$$E_{ini\,(i)} = \frac{1}{P+1} \sum_{j=i-P}^{i} \sum_{m=m_{iniU}}^{m_{iniB}} \sum_{n=n_{iniL}}^{n_{iniR}} b_{(i)\,m,n} : b_{(i)\,m,n} = 1 \,. \tag{3}$$

The averaging values are applied because of uneven distribution of edge elements of objects representing moving vehicles in the binary images. Calculation of the averaging numbers of edge elements is similarly to application of low-pass filters.

The relative value of the averaging number of the edge elements inside the initial detection field in relation to the maximum value of the edge elements within the initial detection field is expressed by the equation.

$$R_{ini\,(i)} = \frac{E_{ini(i)}}{E_{iniMAX}} \,. \tag{4}$$

This relative value of the edge elements is employed for determination of the state of the initial detection field. Similarly, the averaging number $E_{fin(i)}$ of the edge elements inside the final detection field, the maximum value of the edge elements within the finale detection field E_{finMAX}, and its relative value $R_{fin(i)}$ are determined for the current binary image.

The detection field can be in the state "free" or in the state "occupied". The initial detection field changes the free state into occupied state if the relative value of the adjusted number of edge elements $R_{ini(i)}$ inside the initial detection field is greater than the preset threshold value denoted by T_{iniON}. This condition begins to be satisfied for the image with the ordinal number i_{iniON}. The initial detection field changes the occupied state into the free state if the relative value $R_{ini(i)}$ is less than the preset threshold value denoted by T_{iniOFF}. Such condition begins to satisfy the binary image with the ordinal number i_{iniOFF}. Similarly, for the final detection field the threshold values T_{finON}, and T_{finOFF} are preset. These threshold values and the relative value of the number of edge elements $R_{fin(i)}$ are used to determine the state of the final detection field. The state of the final detection field begins to be "occupied" and back "free" for the images with the ordinal numbers i_{finON} and i_{finOFF} respectively.

5   Description of Vehicle Passage

Averaging numbers of edge elements are employed for description of vehicle passages. The vehicle moving through the detection fields change their state. In the beginning both detecting fields are in the free state. At first the vehicle is driving into the initial detection field and changes its state from "free" for the image with the ordinal number equal to $i_{iniON} - 1$ into "occupied" for the image with the ordinal number equal to i_{iniON}. Then the vehicle is leaving the initial detection field and its state changes back from the state "occupied" for the image with the ordinal number equal to $i_{iniOFF} - 1$

into the state "free" for the image with the ordinal number equal to i_{iniOFF}. Analogous changes of the state of the final detection field occur while the vehicle is passing it. These changes are assigned to the pairs of images with the ordinal numbers $i_{\text{finON}} - 1$, i_{finON} and $i_{\text{finOFF}} - 1$, i_{finOFF} respectively.

Vehicles passing the detection fields are labeled by a vehicle number denoted by k. Each vehicle driving through the detection fields is described by the following set of parameters:

$$\left\{ k,\ i_{\text{iniON}(k)},\ i_{\text{iniOFF}(k)},\ i_{\text{finON}(k)},\ i_{\text{finOFF}(k)} \right\} . \tag{5}$$

The sets of parameters describing vehicles passage through the detection fields are applied for determination of the distance between successive vehicles.

6 Distance Between Vehicles

Vehicles following the road are crossing the initial detection field one by one. The frame rate of the applied video camera is denoted by r and expressed in frames per second. A time period between the vehicle which is crossing the initial detection field and the preceding vehicle is given in seconds by the equation:

$$t_{(k)} = \frac{i_{\text{iniON}(k)} - i_{\text{iniON}(k-1)}}{r} . \tag{6}$$

The distance between two consecutive vehicles depends on a time period separated these vehicles and a speed of vehicles. The detection fields are situated at the distance denoted by l. A speed of the k labelled vehicle, driving through the area with the detection fields, can be expressed by:

$$v_{(k)} = \frac{rl}{i_{\text{finON}(k)} - i_{\text{iniON}(k)}} . \tag{7}$$

If the distance between the detection fields is expressed in meters and the frame rate is expressed in frames per second, the vehicle speed is expressed in meters per second. The spatial distance between two consecutive vehicles is equal to the product of the time period between these vehicles and the speed of the preceded vehicle as follows:

$$d_{(k)} = t_{(k)} v_{(k-1)} = l \, \frac{i_{\text{iniON}(k)} - i_{\text{iniON}(k-1)}}{i_{\text{finON}(k-1)} - i_{\text{iniON}(k-1)}} . \tag{8}$$

The detection fields are located at a constant distance, hence the distance between successive vehicles is a function of only image ordinal numbers. The distance between successive vehicles is estimated on the assumption that a speed of both vehicles is constant during measurements. This assumption limits estimation of the spatial distance between vehicles to measured small time periods of the order of seconds. Determination of the time period between vehicles is unlimited.

7   Experimental Results

The frame rate of the video stream obtained at the measuring station is equal to 30 frames per second. Input gray-scale images have intensity resolution of 8 bits per pixel. Each input image is of size 384 × 384 pixels. The detection fields are defined as rectangles. The size of the initial detection field is 80 × 5 pixels. The final detection field is of size 120 × 5 pixels. Figure 7 shows the samples of the input images and binary images presenting two successive vehicles moving through the detection fields. The same vehicles in the further images from the sequence of input images are shown in Fig. 8. Black rectangles indicate the detection fields. In the binary images, the edge elements are indicated by black points.

Fig. 7. Passage of the vehicles through the detection fields (1)

Fig. 8. Passage of the vehicles through the detection fields (2)

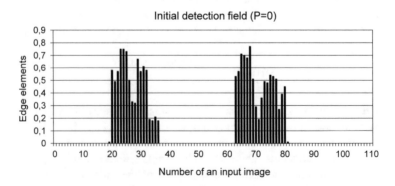

Fig. 9. Relative value of the edge elements in the initial detection field at $P = 0$

For images from the sequence of input images, averaging values of the number of edge elements within detection fields has been calculated without consideration of previous images (the number of previous images is $P = 0$). For the initial detection field, the relative value of the number of the edge elements in individual images shows Fig. 9 and for the final detection field Fig. 10. Averaging values of the number of the edge elements calculated with $P = 0$ are employed for determination of the changes of the detection fields state from "free" into "occupied".

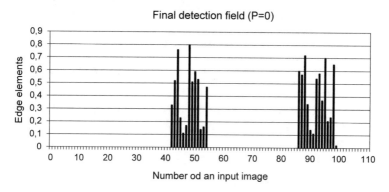

Fig. 10. Relative value of the edge elements in the final detection field at $P = 0$

The threshold value for the changes from the free state into the occupied state for both detection fields has been preset to 0.2. The time period between two considered vehicles is determined as 1.43 s. The detection fields are located in the distance equal to 10 m, hence the distance between two considered vehicles is estimate at 19.6 m.

For images from the sequence of input images, averaging values of the number of edge elements in the area of the detection fields also has been calculated with consideration of 4 previous images (the number of previous images is $P = 4$). The relative averaging values of the number of the edge elements in the individual images determined for the initial and final detection fields are shown in Figs. 11 and 12 respectively.

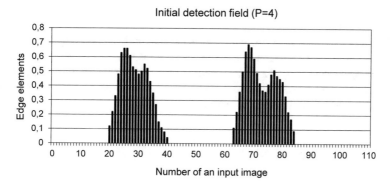

Fig. 11. Relative value of the edge elements in the initial detection field at $P = 4$

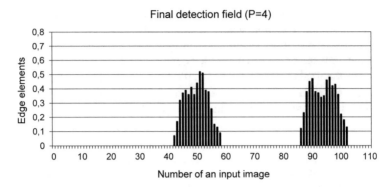

Fig. 12. Relative value of the edge elements in the final detection field at $P = 4$

Averaging values of the number of the edge elements calculated with $P = 4$ are utilized for determination of the back changes of the of the state of the detection fields from the occupied state into the free state.

8   Properties of the Method

In the proposed method of distance estimation, the geometry of a measurement station is not significant, but location of a video camera should allow to obtain a correct video stream suitable for further processing. High parameters of a video camera are not required, and practically each video camera is suitable for acquisition of vision data at a measuring station.

In the method of distance estimation between vehicles, determination of time distance has no any assumption limits. Estimation of spatial distance is carried out with the assumption of constant speed of vehicles because a vehicle speed is calculated only between detection fields. Threshold values utilized for determination of the state of detection fields can be adapted to a size and resolution of processed images, thus size and resolution of applied images does not limit usefulness of the method. Real distance between detection fields depends on definition of the detection fields for a sequence of images, and is a factor of proportionality in equation form describing a spatial distance between vehicles.

Efficiency of video-based distance estimation depends on efficiency of detection of vehicles. Video-based detection of distance between vehicles is similarly susceptible to disturbances, especially atmospheric disturbances. Validation of the obtained results are carried out with the consideration of data from the camera inner clock and geometric properties of the measuring station.

9   Conclusions

A distance between vehicles in a vehicle stream can be estimated with the use of vision data. In the presented method, conversion of input images into binary images is applied. Conversion into binary images is performed with the use of small gradients determined in input images. Location of the edge elements in the obtained binary images is in accordance with scene components of input images.

The proposed method of estimation of the distance between successive vehicles in the stream of vehicles is simply and does not required complicated operations. The distance between successive vehicles is estimated with the use of appropriate operation performed on ordinal numbers of images from the sequence of input images. Estimation of the distance between vehicles in streams of vehicles is intended for measurements of traffic parameters and for application in systems of traffic surveillance.

References

1. Coifman, B., Beymer, D., McLauchlan, P., Malik, J.: A real-time vision system for vehicle tracking and traffic surveillance. Transp. Res. Part C **6**, 271–288 (1998)
2. Fernandez-Caballero, A., Gomez, F.J., Lopez-Lopez, J.: Road traffic monitoring by knowledge-driven static and dynamic image analysis. Expert Syst. Appl. **35**, 701–719 (2008)
3. Kamijo, S., Matsushita, Y., Ikeuchi, K., Sakauchi, M.: Traffic monitoring and accident detection at intersections. IEEE Trans. Intell. Transp. Syst. **1**(2), 108–118 (2000)
4. Gupte, S., Masoud, O., Martin, R.F.K., Papanikolopoulos, N.P.: Detection and classification of vehicles. IEEE Trans. Intell. Transp. Syst. **3**(1), 37–47 (2002)
5. Hsieh, J., Yu, S.H., Chen, Y.S., Hu, W.F.: Automatic traffic surveillance system for vehicle tracking and classification. IEEE Trans. Intell. Transp. Syst. **7**(2), 175–187 (2006)
6. Cucchiara, R., Piccardi, M., Mello, P.: Image analysis and rule-based reasoning for a traffic monitoring system. IEEE Trans. Intell. Trans. Syst. **1**(2), 119–130 (2000)
7. Mithun, N.C., Rashid, N.U., Rahman, S.M.M.: Detection and classification of vehicles from video using multiple time-spatial images. IEEE Trans. Intell. Transp. Syst. **13**(3), 1215 1225 (2012)
8. Stauffer, C., Grimson, W.E.L.: Adaptive background mixture model for real-time tracking. Proc. IEEE Comput. Soc. Conf. Comput. Vis. Pattern Recogn. **2**, 246–252 (1999)
9. Dailey, D.J., Cathey, F.W., Pumrin, S.: An algorithm to estimate mean traffic speed using uncalibrated cameras. IEEE Trans. Intell. Transp. Syst. **1**(2), 98–107 (2000)
10. Rahim, H.A., Sheikh, U.U., Ahmad, R.B., Zain, A.S.M., Ariffin, W.N.F.: Vehicle speed detection using frame differencing for smart surveillance system. 10th International Conference on Information Science. Signal Processing and their Applications (ISSPA), pp. 630–633. IEEE, Kuala Lumpur (2010)
11. Czapla, Z.: Point image representation for efficient detection of vehicles. In: Burduk, R., Jackowski, K., Kurzyński, M., Woźniak, M., Żołnierek, A. (eds.) Proceedings of the 9th International Conference on Computer Recognition Systems CORES 2015. AISC, vol. 403, pp. 691–700. Springer, Cham (2016). https://doi.org/10.1007/978-3-319-26227-7_65
12. Czapla, Z.: Vehicle speed estimation with the use of gradient-based image conversion into binary form. Signal Processing. Algorithms, Architectures, Arrangements, and Applications (SPA), Conference Proceedings, pp. 213–216. Semantic Scholar, Seattle (2017)

Safe IT System to Assist in Control of Safe Locker Transport Loads

Paweł Buchwald[1(✉)] and Artur Anus[2]

[1] WSB University, Dąbrowa Górnicza, Poland
pbuchwald@wsb.edu.pl
[2] Iconity sp. z. o. o., Katowice, Poland
artur.anus@iconity.pl

Abstract. The article presents main goals and architectural implementation of the system for supporting the control of transport cargo, as well as preliminary results of tests in environmental conditions, which are identical as the conditions of natural exploitation. The aim of the article is the publication of research results collected during the design, implementation and operation of the system. The authors show proposed solutions to problems that appeared during the creation of the system by the research and development team. The article presents also the completed development work for security improvement through integration with modern blockchain-based data processing systems. The research on the presented IT system to support the control of the Safe Locker transport loads was made as part of the project.

Keywords: Ethereum · Block register · Electronic seals · Transport security

1 Introduction and Literature Review

The need to control transport loads is an important area supporting the activities of many enterprises. This necessity results from the requirements of protection of the cargo itself against theft, but also from the functionality of the supply chain tracking. Enterprises are trying to minimize their inventory, which is dictated by the introduction of Industry 4.0. Therefore, effective delivery of transport cargo becomes a key aspect in the correct operation of many enterprises. The Safe Locker system meets these needs by providing the possibility of electronic monitoring of access to the ability to track and secure transport loads, shed information on events related to unauthorized access, and direct current tracking of cargo. The article presents the construction and use tests of the Safe Locker system, which was implemented on the basis of commercially available devices used to build IoT systems, focusing on the concepts of IT architecture and the results of tests performed during ongoing operation. The experience of the research team shows that it is possible to implement such a system based on devices used in IoT systems that automate data acquisition. Due to the direct dependence of the system operation possibilities on the quality of GSM and GPRS signals, the article presents quality tests of distributed data in the context of dependence on the GPS and GPRS signal efficiency.

© Springer Nature Switzerland AG 2020
E. Macioszek and G. Sierpiński (Eds.): Modern Traffic Engineering in the System Approach to the Development of Traffic Networks, AISC 1083, pp. 296–305, 2020.
https://doi.org/10.1007/978-3-030-34069-8_23

The concept of the Internet of Things was initiated in the areas of goods tracking. For the first time, this concept was used by Kevin Ashton in 1999. Currently, the development of this concept has led to significant progress in technical solutions in this area [1, 2]. Despite the introduction of network communication standards dedicated to IoT systems, mobile phone networks are still used in this area due to its popularity and coverage of its wide area [3]. In the implementation of IoT systems, energy efficiency is also an important aspect, which determines the system's ability to operate in the field for a longer period [4]. In IoT systems, an important aspect when implementing communication between devices is to maintain the security of data acquisition. This is difficult when using the Ad Hoc network [5]. The implementation concept of the Safe Locker system envisages for this reason the use of solutions made available in infrastructure networks, while maintaining the energy efficiency of position measurement devices and cargo access control. IoT systems increasingly use decentralized architecture offered by systems built on the basis of Blockchain [6]. Also in the Safe Locker system, an IoT device management module based on the Ethereum system was used, which allowed the use of decentralized architecture. Due to the need to integrate the system with traditional data processing platforms (for example operating within ERP systems), it was also necessary to use the HA architecture eliminating a single point of failure in intermediary modules [7].

2 Functionality of Implemented Control System for Assistance in Transport Loads

Basic system functionality includes monitoring and notifying of unauthorized access to loads, and at the same time does not involve a physical protection - that is, the system is designed to notify of occurrence of a specific, undesirable situation, and simultaneously not aspiring to the role that might prevent such from occurring. Thus, for this reason the critical parameter for the system is alarm messages propagation rate, which determines its architecture. Basic functionality (intrusion detection) is backed up by aggregation and transmission of additional data, such as location, power supply status, signal strength etc., which contributes not only to system's intrusion detection of load integrity, but also to its condition and location monitoring.

3 Project and Implementation of the System Combined with Building the Apparatus Used to Monitor Transport Loads

The system architecture is based on use of radio technologies for transmission purposes and fiber optic technology for security (Fig. 1). Safeguarding component (called a 'seal') comprises a fiber optic ring threaded through secured elements (discharge valves in wagons, transport containers hoops etc.), through which random data is transmitted. In the event of interruption in loop continuity or unintentional attempt to interfere with transmitted data (e.g. by opening the cover, which changes the degree of lighting and disrupts transmission) such condition is detected as a trial to interfere with load integrity and the alarm is activated, whereas notification of it is immediately sent to server by means of GSM network. Even in case of lack of interruption, seals send data in cycles on its location and condition, including e.g. GSM signal strength, battery charge level, temperature inside the casing, condition of the casing (whether it has not been opened) etc., which enables to monitor both the apparatus and indirectly the train set. In addition, the system is supplemented with mobile application, which using the RFID technology allows an activation of the seal in the system on mounting the protection, as well as its deactivation before it is disassembled - which limits alarm messages solely to situations, where such information corresponds to actuality and is unrelated to installation/dismantling.

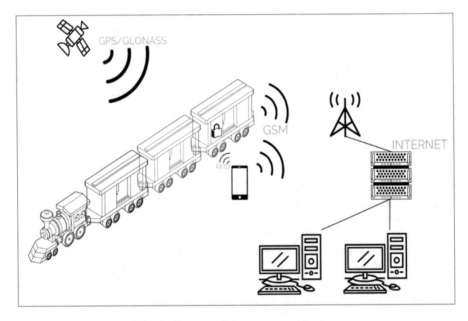

Fig. 1. Concept of the system architecture

Transport sets are logically mapped on server side, where data supplied by apparatus are too collected. By providing access of suitable API, the server allows to apply client terminals of diversified functionality - which enables to track down the train route in a graphic way or notify of breakdowns and extraordinary situations, depending on client's rights and requests. The safeguarding apparatus is composed of three main modules - power-supply module, management module and message module. The first of them is in charge of supporting power supply, provided by configurable sets of lithium-filtered or lithium-polymer battery packs. Voltage is adequately stabilized and filtered, then fed into voltage rails. In order to secure systems, the protective device monitors the charge status, and if necessary notifies of batteries running out and shifting into energy efficient mode.

The management module is the core of the apparatus and coordinates operation of the two remaining and operates a fiber optic loop. The sending-receiving system integrated within the module works in infrared band and consists of optical sensor receiver, photosensitive elements and such conditioning the signal.

Random data is transmitted through loop using binary ASCII code representations, whereas in case of inconsistencies of data received with the ones sent the procedure notifying of alarm is activated. Data intended for transmission via message module are conveyed to it using the Universal Serial Bus. The message module alone may be of integrated construction (GPS and GSM modules all in one) or of split design, however its functional scope (read-out of position using satellite navigation and communication by means of data transmission via GSM network) remains unaltered.

4 Efficacy Tests of Loads Data Acquisition System Performance Run in Natural Operating Conditions

Suggested solution underwent a pilot-run process in real conditions. Within its frameworks railway routes of diverse nature (urban, rural, mixed, at high and low speeds etc.) were travelled, with cards from various mobile phone companies and in different supply configurations. In addition, some items were equipped too with an option to record data in SD card, which enabled to detect possible troubles with communication via GSM. Exemplary results of pilot run on route from Czechowice-Dziedzice to Zduńska Wola Karsznice (railway lines no. 93, 138, 161 and 131) are presented in Figs. 2, 3 and 4. As may be seen in the unit, there occurs a progressive fall in supply voltage, which is correspondent to discharge curves for applied lithium-ion technology [8]. With discharge rate retained, the apparatus could still function in an uninterrupted way for three days in succession, before it reached the warning level (6,4 V, at nominal and secure supply voltage of 5,8 V for a pair of batteries) and shifted into energy efficient mode.

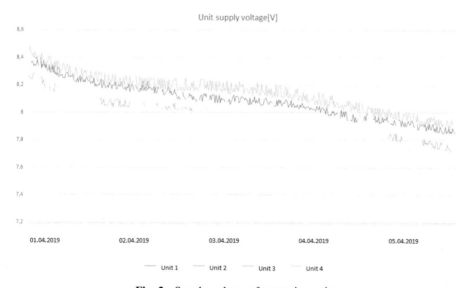

Fig. 2. Supply voltage of protection units

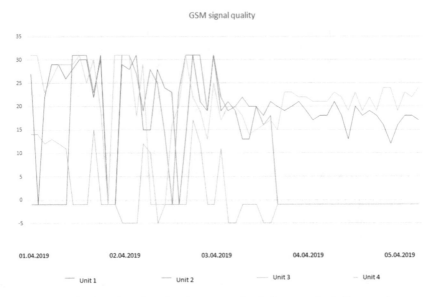

Fig. 3. Signal quality of various networks during stretch of pilot travel

In contrast, as it is shown in Fig. 3, offered signal quality significantly differs depending on mobile phone company - disparities in signal quality (indicated in an abstract unit received by means of standard AT "AT+CSQ" command) may even reach up to a dozen units or so, not to mention that particular networks may be utterly beyond reach, while others are characterized by a satisfactory signal quality. In turn, GPS signal does not cause such problems - most often no report of the apparatus location is not caused by lack of GPS signal, but impossibility to report in relation to GSM disconnection.

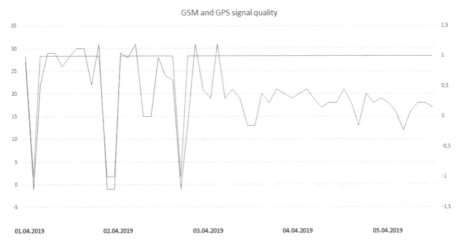

Fig. 4. Signal quality of GPS and GSM for a single apparatus

5 Capabilities of Integrating the System with Ethereum Platform to Safely Manage the Infrastructure of Distributed Devices

Blockchain is one of solutions classified within technologies of dispersed registers. These concepts are often incorrectly treated as interchangeable. Dispersed register is a distributed database, in which each of nodes owns a replicated copy of entirety of data. Each of nodes updates data independently, and the system as a whole is responsible for their synchronizations. The principal assumption of the system based on dispersed registers is lack of central computer - a server that manages access to data. Correctness of updates is verified by node groups, which confirm correctness and security of transactions by means of dedicated methods of reaching a consensus.

From the perspective of analysis of IT systems blockchain is the database of decentralized architecture, in which data storage uses blocks tied together to create chains. Linking is carried out based on complex cryptographic algorithms in terms of calculation, which ensures cohesion and security of the system by virtue of attempts at unauthorized modifications. The principal difference between a typical distributed IT system and the system based on blockchain is a default trust in nodes in case of traditional dispersed IT systems. blockchain system is characterized by the thing that trust in all nodes ought not to be determined by default [8, 9].

It can only be assumed that most nodes (over 50%) are certified and trusted appliances. One of features of distributed blockchain database is its appearance in many identical copies. Each copy contains a set of data in the form of tied-up blocks. Blockchain-based system ensures irreversibility of records done in dispersed database. Attempt to change a single block entails necessity of modification of the whole chain, which from the point of view of essential calculating power is unusually difficult to accomplish. Technology is based on point-point networks, without central computers, which does not lead to emergence of a single element of failure in the form of server storing data. System of dispersed registers is much more restricted in terms of options of conduction of operations on data in comparison to distributed databases. In the event of application of systems based on blockchains, realisation of a single transaction of data modification is much slower. Application of systems based on Blockchain is a noteworthy solution when the overriding task is to assure safety, not time effectiveness of processed data [10].

In the Ethereum network like in case of traditional cryptocurrency systems, next blocks are created in extraction process. The block shall record the list of transactions, which were affected since the time of realisation of the last transaction recorded in the previous block and value of hash function that represents new tree status on realisation of confirmed transactions. The originator of a newly created block is awarded by the system for its acquisition, so called mining the block. Blocks are linked together to form a characteristic chain (blockchain). Conception of data stored in such block is shown in Fig. 5.

At present, Ethereum platform may be perceived as the blockchain system, which operates a high-level java script language used for creating built-in elements of business logic. Attention should also be drawn to the fact that Ethereum platform introduces the virtual machine into activation of the code of created distributed applications. Virtual machine operates on many dispersed nodes and is based on assuring safety of activated code by means of algorithmic methods of reaching a consensus by nodes belonging to network. Ethereum Virtual Machine may change its status under influence of released actions that allow to activate the function implemented in java script language, which is referred to as contract.

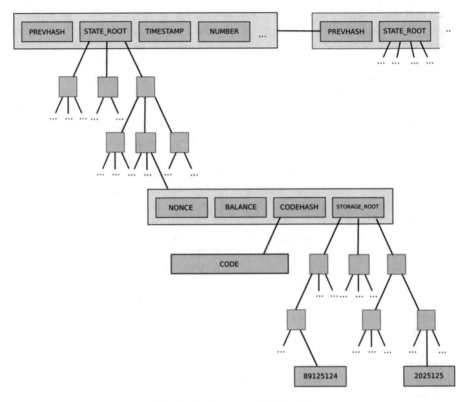

Fig. 5. Architecture of blockchain

In dispersed applications created with the help of Ethereum platform, the contract serves the purpose of realization of four basic functions:

- storage of status of useful data in memory, which may be essential from the point of view of other contracts or external applications. An example of such information may be the contract that might stand for portfolio state in ETH, but also information relating to ownership of a given material stock,
- performance of conditional operations provided that additional requirements are fulfilled. It is so-called forwarding contract. Such type of contract allows to transmit messages to a target object if extra conditions are fulfilled, which may be verified by means of fragment of the code. An example may be realisation of conditional ETH currency transfer if information of ownership of the stock or access to it is modified.

Conditional operations may also be used to realise functionality of multilateral signature. If information is confirmed, e.g. with at least three out of five chosen private keys the next contract will be possible to be set in motion,

- management of status of other contracts. This sort of contract may be useful for performance of financial agreement controlled by external entities or with insurance transactions. This type of contract allows to define a target object later. For instance, implementation of tender functionality is possible by its means. EOA account that will offer the highest value shall be the target object of the contract, which transmits information of ownership of stock,
- fulfills functions of a supplier of modules and libraries for other contracts.

Interaction between contracts is held on the basis of exchange of messages. Sending the message may be initiated in a programmed way on the contract level or from the Ethereum user's account when carrying out a payment transaction by means of cryptocurrency [11].

6   Increasing the Security of the System Using Private APN Networks

The possibilities of data acquisition in the Internet allow you to send measurement data by reading additional authorization information from an Ethereum-based system. One of the parameters enabling the proper operation of the systems is the quality of the GSM/LTE signal. The quality tests of this parameter were presented in the previous sub-chapter. The quality of the signal depends on many factors, including the distance from the BTS stations. For this reason, the correct operation of the system may be determined by the selection of SIM cards of the respective operators. SIM cards designed for packet data transmission allow the device to be assigned to the APN packet network. APN networks available on the market of operators enable the provision of private and public services. Public networks are open and generally offer device addressing with dynamic IP addresses. In telemetry applications, it is convenient to use static addresses that allow direct communication between devices of the monitoring system. The use of static IP addresses in public networks is associated with additional risk and greater vulnerability to attacks. Lack of security in the form of a closed teletransmission structure exposes the device to receive data from external sources. DOS (Denial of services) attacks are possible, and therefore private APN networks are often used in telemetric systems. Such networks allow separating a group of telemetry devices, servers, and workstations, to which a special level of security should be applied.

The architecture of the presented solution, however, assumes the necessity to propose communication with the Ethereum network. This is possible using the APN private network architecture with the simultaneous option of using intermediary stations to control external network traffic. This architecture allows packet control using access control lists, which minimizes network traffic. It is also possible to control access to selected services by blocking appropriate ports for TCP/IP and UDP.

The presented architecture has some limitations related to the existence of an intermediary module in data transmission between private and public networks. Such a module constitutes a single point of failure, and if there is a need to send many messages, it may constitute a bottleneck of the system. The solution of this problem is implemented in the case of HA systems design through resource redundancy.

Information from the Safe Locker system can be used by other IT systems. Providing data on cargo geolocation is of great value in the area of logistic business services. There is a need to distribute data from the Safe Locker system to external ERP systems. In order to meet the requirements of a high level of system security, as a whole, data acquisition can be carried out via a VPN network. The majority of the service providers providing APN network services available on the market also offer services of matching dedicated VPN tunnels from private APN networks to the selected client infrastructure.

An illustrative system architecture that meets all the above-mentioned security requirements and at the same time allows communication with public IT services available on the Internet is shown in Fig. 6.

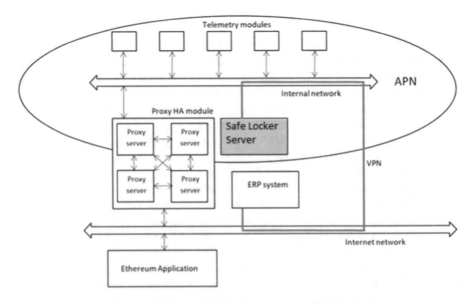

Fig. 6. Architecture of distributed system Safe Locker

7 Conclusion and Further Development Work

One of additional aspects of use of Ethereum network is facilitation of performance of tasks related to electronic payments. Dispersed register of Ethereum platform may store information of status of chosen nodes of loads control system along with financial and accounting information relating to exploitation of appliances. Development work on presented system shall pertain to further use of Ethereum system for improvement to safety of data, as well as storage of information on quality of distributed surveys. This will allow to use a redundant channel of data acquisition in the form of communication channel between Ethereum platform and the system of access control. Such approach offers an opportunity to ensure verification of authenticity of sent messages via system of safe cryptographic keys, while a redundant platform of data exchange may be used for exchange of public keys [8, 12]. Ethereum platform is proven in cryptocurrency applications and may be recognized as the system that complies with requirements of messages acquisition safety.

References

1. Bandyopadhyay, D., Sen, J.: The internet of things - applications and challenges in technology and standardization. Wirel. Pers. Commun. **58**(1), 49–69 (2011)
2. Atzori, L., Iera, A., Morabito, G.: The Internet of Things a survey. Comput. Netw. **54**, 2787–2805 (2010)
3. Aloi, G., Caliciuri, G., Fortino, G., Gravina, R., Pace, P., Russo, W., Savaglio, C.: Enabling IoT interoperability through opportunistic smartphone-based mobile gateways. J. Netw. Comput. Appl. **81**, 74–84 (2017)
4. Chen, J., Shen, H.: Meleach-L: More energy-efficient leach for large-scale WSNs. In: 2008 4th Internationalk Conference on Wireless Communications, Networking and Mobile Computing, pp. 1–4. IEEE Press (2008)
5. Belloa, O., Zeadally, S., Badrac, M.: Network layer inter-operation of Device-to-Device communication technologies in Internet of Things (IoT). J. Ad Hoc Netw. **57**(C), 52–62 (2017)
6. Huh, S., Kim, S.: Managing IoT devices using blockchain platform. In: 19th International Conference on Advanced Communication Technology, Bongpyeong, pp. 464–467. IEEE Press (2017)
7. Aazam, M., Khan, I., Alsaffar, A., Huh, N.: Cloud of things: integrating internet of things and cloud computing and the issues involved. In: Proceedings of 2015 11th International Bhurban Conference on Applied Science & Technology, Alaska, pp. 414–419. IEEE Press (2014)
8. Li-Ion & LiPoly Batteries Discharge. https://learn.adafruit.com/li-ion-and-lipoly-batteries/voltages
9. Technical documentation of platform Ethereum. https://www.ethereum.org
10. Wood, G.: Ethereum: a secure decentralized generalized transaction ledger. https://gavwood.com/paper.pdf
11. Nowakowski, M.: Ethereum Smart Contracts: Security Vulnerabilities and Security Tools. Norwegian University of Science and Technology. https://brage.bibsys.no/xmlui/bitstream/handle/11250/2479191/18400_FULLTEXT.pdf
12. Buchwald, P., Rostański, M., Mączka, K.: Network steganography method for users identity confirmation in web applications. Theor. Appl. Inform. **26**, 177–187 (2014)

Author Index

© Springer Nature Switzerland AG 2020
E. Macioszek and G. Sierpiński (Eds.): Modern Traffic Engineering in the System Approach
to the Development of Traffic Networks, AISC 1083, pp. 307–308, 2020.
https://doi.org/10.1007/978-3-030-34069-8

Printed in the United States
By Bookmasters